AF892625

Functional Marine Biomaterials

Related titles

Drug Device Combination Products
(ISBN: 9781845694708)

Natural Based Polymers for Biomedical Applications
(ISBN: 9781845692643)

Biomedical Hydrogels
(ISBN: 9781845695903)

Tissue Engineering Using Ceramics and Polymers
(ISBN: 9781845691769)

Electrospinning for Tissue Regeneration
(ISBN: 9781845697419)

Woodhead Publishing Series in Biomaterials: Number 100

Functional Marine Biomaterials

Properties and Applications

Edited by

Se-Kwon Kim

ELSEVIER

AMSTERDAM • BOSTON • CAMBRIDGE • HEIDELBERG
LONDON • NEW YORK • OXFORD • PARIS • SAN DIEGO
SAN FRANCISCO • SINGAPORE • SYDNEY • TOKYO
Woodhead Publishing is an imprint of Elsevier

WP
WOODHEAD PUBLISHING

Woodhead Publishing Limited is an imprint of Elsevier
80 High Street, Sawston, Cambridge, CB22 3HJ, UK
225 Wyman Street, Waltham, MA 02451, USA
Langford Lane, Kidlington, OX5 1GB, UK

Copyright © 2015 Elsevier Ltd. All rights reserved.

No part of this publication may be reproduced or transmitted in any form or by any means, electronic or mechanical, including photocopying, recording, or any information storage and retrieval system, without permission in writing from the publisher. Details on how to seek permission, further information about the Publisher's permissions policies and our arrangements with organizations such as the Copyright Clearance Center and the Copyright Licensing Agency, can be found at our website: www.elsevier.com/permissions.

This book and the individual contributions contained in it are protected under copyright by the Publisher (other than as may be noted herein).

Notices

Knowledge and best practice in this field are constantly changing. As new research and experience broaden our understanding, changes in research methods, professional practices, or medical treatment may become necessary.

Practitioners and researchers must always rely on their own experience and knowledge in evaluating and using any information, methods, compounds, or experiments described herein. In using such information or methods they should be mindful of their own safety and the safety of others, including parties for whom they have a professional responsibility.

To the fullest extent of the law, neither the Publisher nor the authors, contributors, or editors, assume any liability for any injury and/or damage to persons or property as a matter of products liability, negligence or otherwise, or from any use or operation of any methods, products, instructions, or ideas contained in the material herein.

ISBN: 978-1-78242-086-6 (print)
ISBN: 978-1-78242-094-1 (online)

British Library Cataloguing in Publication Data

A catalogue record for this book is available from the British Library

Library of Congress Control Number: 2015942338

For Information on all Woodhead Publishing publications visit our website at http://store.elsevier.com/

Working together to grow libraries in developing countries

www.elsevier.com • www.bookaid.org

Contents

List of contributors

Toshihiro Akaike Tokyo Institute of Technology, Yokohama, Japan

B. Ben-Nissan University of Technology, Sydney, Australia

Chong-Su Cho Seoul National University, Seoul, Republic of Korea

Yun-Jaie Choi Seoul National University, Seoul, Republic of Korea

S. Colliec-Jouault Laboratory of Biotechnology and Marine Molecules, Nantes Cedex, France

E. Giuliani University of Modena and Reggio Emilia, Modena, Italy

Tao Jiang Seoul National University, Seoul, Republic of Korea

S.-K. Kim Marine Bioprocess Research Center, Pukyong National University, Busan, South Korea

B. Lowe Pukyong National University, Busan, South Korea

Werner E.G. Müller University Medical Center of the Johannes Gutenberg University Mainz, Mainz, Germany

R. Parravicini University of Modena and Reggio Emilia, Modena, Italy

V. Parravicini University of Perpignan, Perpignan, France

R.L. Reis University of Minho, Taipas, Guimarães, Portugal and ICVS/3B's – PT Government Associate Laboratory, Braga/Guimarães, Portugal

Heinz C. Schröder University Medical Center of the Johannes Gutenberg University Mainz, Mainz, Germany

S. Şenel Hacettepe University, Ankara, Turkey

C.P. Sharma Sree Chitra Tirunal Institute for Medical Science and Technology, Kerala, India

T.H. Silva University of Minho, Taipas, Guimarães, Portugal and ICVS/3B's – PT Government Associate Laboratory, Braga/Guimarães, Portugal

Bijay Singh Seoul National University, Seoul, Republic of Korea

J. Venkatesan Pukyong National University, Busan, Korea

S.P. Victor Sree Chitra Tirunal Institute for Medical Science and Technology, Kerala, India

Xiaohong Wang University Medical Center of the Johannes Gutenberg University Mainz, Mainz, Germany

Woodhead Publishing Series in Biomaterials

1 **Sterilisation of tissues using ionising radiations**
Edited by J.F. Kennedy, G.O. Phillips and P.A. Williams
2 **Surfaces and interfaces for biomaterials**
Edited by P. Vadgama
3 **Molecular interfacial phenomena of polymers and biopolymers**
Edited by C. Chen
4 **Biomaterials, artificial organs and tissue engineering**
Edited by L. Hench and J. Jones
5 **Medical modelling**
R. Bibb
6 **Artificial cells, cell engineering and therapy**
Edited by S. Prakash
7 **Biomedical polymers**
Edited by M. Jenkins
8 **Tissue engineering using ceramics and polymers**
Edited by A.R. Boccaccini and J. Gough
9 **Bioceramics and their clinical applications**
Edited by T. Kokubo
10 **Dental biomaterials**
Edited by R.V. Curtis and T.F. Watson
11 **Joint replacement technology**
Edited by P.A. Revell
12 **Natural-based polymers for biomedical applications**
Edited by R.L. Reiss et al.
13 **Degradation rate of bioresorbable materials**
Edited by F.J. Buchanan
14 **Orthopaedic bone cements**
Edited by S. Deb
15 **Shape memory alloys for biomedical applications**
Edited by T. Yoneyama and S. Miyazaki
16 **Cellular response to biomaterials**
Edited by L. Di Silvio
17 **Biomaterials for treating skin loss**
Edited by D.P. Orgill and C. Blanco
18 **Biomaterials and tissue engineering in urology**
Edited by J. Denstedt and A. Atala
19 **Materials science for dentistry**
B.W. Darvell
20 **Bone repair biomaterials**
Edited by J.A. Planell, S.M. Best, D. Lacroix and A. Merolli

21 **Biomedical composites**
Edited by L. Ambrosio
22 **Drug–device combination products**
Edited by A. Lewis
23 **Biomaterials and regenerative medicine in ophthalmology**
Edited by T.V. Chirila
24 **Regenerative medicine and biomaterials for the repair of connective tissues**
Edited by C. Archer and J. Ralphs
25 **Metals for biomedical devices**
Edited by M. Ninomi
26 **Biointegration of medical implant materials: Science and design**
Edited by C.P. Sharma
27 **Biomaterials and devices for the circulatory system**
Edited by T. Gourlay and R. Black
28 **Surface modification of biomaterials: Methods analysis and applications**
Edited by R. Williams
29 **Biomaterials for artificial organs**
Edited by M. Lysaght and T. Webster
30 **Injectable biomaterials: Science and applications**
Edited by B. Vernon
31 **Biomedical hydrogels: Biochemistry, manufacture and medical applications**
Edited by S. Rimmer
32 **Preprosthetic and maxillofacial surgery: Biomaterials, bone grafting and tissue engineering**
Edited by J. Ferri and E. Hunziker
33 **Bioactive materials in medicine: Design and applications**
Edited by X. Zhao, J.M. Courtney and H. Qian
34 **Advanced wound repair therapies**
Edited by D. Farrar
35 **Electrospinning for tissue regeneration**
Edited by L. Bosworth and S. Downes
36 **Bioactive glasses: Materials, properties and applications**
Edited by H.O. Ylänen
37 **Coatings for biomedical applications**
Edited by M. Driver
38 **Progenitor and stem cell technologies and therapies**
Edited by A. Atala
39 **Biomaterials for spinal surgery**
Edited by L. Ambrosio and E. Tanner
40 **Minimized cardiopulmonary bypass techniques and technologies**
Edited by T. Gourlay and S. Gunaydin
41 **Wear of orthopaedic implants and artificial joints**
Edited by S. Affatato
42 **Biomaterials in plastic surgery: Breast implants**
Edited by W. Peters, H. Brandon, K.L. Jerina, C. Wolf and V.L. Young
43 **MEMS for biomedical applications**
Edited by S. Bhansali and A. Vasudev
44 **Durability and reliability of medical polymers**
Edited by M. Jenkins and A. Stamboulis

45 **Biosensors for medical applications**
Edited by S. Higson
46 **Sterilisation of biomaterials and medical devices**
Edited by S. Lerouge and A. Simmons
47 **The hip resurfacing handbook: A practical guide to the use and management of modern hip resurfacings**
Edited by K. De Smet, P. Campbell and C. Van Der Straeten
48 **Developments in tissue engineered and regenerative medicine products**
J. Basu and J.W. Ludlow
49 **Nanomedicine: Technologies and applications**
Edited by T.J. Webster
50 **Biocompatibility and performance of medical devices**
Edited by J.-P. Boutrand
51 **Medical robotics: Minimally invasive surgery**
Edited by P. Gomes
52 **Implantable sensor systems for medical applications**
Edited by A. Inmann and D. Hodgins
53 **Non-metallic biomaterials for tooth repair and replacement**
Edited by P. Vallittu
54 **Joining and assembly of medical materials and devices**
Edited by Y. (Norman) Zhou and M.D. Breyen
55 **Diamond-based materials for biomedical applications**
Edited by R. Narayan
56 **Nanomaterials in tissue engineering: Fabrication and applications**
Edited by A.K. Gaharwar, S. Sant, M.J. Hancock and S.A. Hacking
57 **Biomimetic biomaterials: Structure and applications**
Edited by A.J. Ruys
58 **Standardisation in cell and tissue engineering: Methods and protocols**
Edited by V. Salih
59 **Inhaler devices: Fundamentals, design and drug delivery**
Edited by P. Prokopovich
60 **Bio-tribocorrosion in biomaterials and medical implants**
Edited by Y. Yan
61 **Microfluidic devices for biomedical applications**
Edited by X.-J. James Li and Y. Zhou
62 **Decontamination in hospitals and healthcare**
Edited by J.T. Walker
63 **Biomedical imaging: Applications and advances**
Edited by P. Morris
64 **Characterization of biomaterials**
Edited by M. Jaffe, W. Hammond, P. Tolias and T. Arinzeh
65 **Biomaterials and medical tribology**
Edited by J. Paolo Davim
66 **Biomaterials for cancer therapeutics: Diagnosis, prevention and therapy**
Edited by K. Park
67 **New functional biomaterials for medicine and healthcare**
E.P. Ivanova, K. Bazaka and R.J. Crawford
68 **Porous silicon for biomedical applications**
Edited by H.A. Santos

69 **A practical approach to spinal trauma**
Edited by H.N. Bajaj and S. Katoch
70 **Rapid prototyping of biomaterials: Principles and applications**
Edited by R. Narayan
71 **Cardiac regeneration and repair: Volume 1: Pathology and therapies**
Edited by R.-K. Li and R.D. Weisel
72 **Cardiac regeneration and repair: Volume 2: Biomaterials and tissue engineering**
Edited by R.-K. Li and R.D. Weisel
73 **Semiconducting silicon nanowires for biomedical applications**
Edited by J.L. Coffer
74 **Silk biomaterials for tissue engineering and regenerative medicine**
Edited by S. Kundu
75 **Biomaterials for bone regeneration: Novel techniques and applications**
Edited by P. Dubruel and S. Van Vlierberghe
76 **Biomedical foams for tissue engineering applications**
Edited by P. Netti
77 **Precious metals for biomedical applications**
Edited by N. Baltzer and T. Copponnex
78 **Bone substitute biomaterials**
Edited by K. Mallick
79 **Regulatory affairs for biomaterials and medical devices**
Edited by S.F. Amato and R. Ezzell
80 **Joint replacement technology. Second edition**
Edited by P.A. Revell
81 **Computational modelling of biomechanics and biotribology in the musculoskeletal system: Biomaterials and tissues**
Edited by Z. Jin
82 **Biophotonics for medical applications**
Edited by I. Meglinski
83 **Modelling degradation of bioresorbable polymeric medical devices**
Edited by J. Pan
84 **Perspectives in total hip arthroplasty: Advances in biomaterials and their tribological interactions**
S. Affatato
85 **Tissue engineering using ceramics and polymers. Second edition**
Edited by A.R. Boccaccini and P.X. Ma
86 **Biomaterials and medical-device associated infections**
Edited by L. Barnes and I.R. Cooper
87 **Surgical techniques in total knee arthroplasty (TKA) and alternative procedures**
Edited by S. Affatato
88 **Lanthanide oxide nanoparticles for molecular imaging and therapeutics**
G.H. Lee
89 **Surface modification of magnesium and its alloys for biomedical applications: Volume 1: Biological interactions, mechanical properties and testing**
Edited by T.S.N. Sankara Narayanan, I.S. Park and M.H. Lee
90 **Surface modification of magnesium and its alloys for biomedical applications: Volume 2: Modification and coating techniques**
Edited by T.S.N. Sankara Narayanan, I.S. Park and M.H. Lee

91 **Medical modelling: the application of advanced design and rapid prototyping techniques in medicine. Second edition**
Edited by R. Bibb, D. Eggbeer and A. Paterson

92 **Switchable and responsive surfaces and materials for biomedical applications**
Edited by Z. Zhang

93 **Biomedical textiles for orthopaedic and surgical applications: fundamentals, applications and tissue engineering**
Edited by T. Blair

94 **Surface coating and modification of metallic biomaterials**
Edited by C. Wen

95 **Hydroxyapatite (HAP) for biomedical applications**
Edited by M. Mucalo

96 **Implantable neuroprostheses for restoring function**
Edited by K. Kilgore

97 **Shape memory polymers for biomedical applications**
Edited by L. Yahia

98 **Regenerative engineering of musculoskeletal tissues and interfaces**
Edited by S.P. Nukavarapu, J.W. Freeman and C.T. Laurencin

99 **Advances in cardiac imaging: techniques and applications**
Edited by K. Nieman, O. Gaemperli, P. Lancellotti and S. Plein

100 **Functional Marine Biomaterials: Properties and Applications**
Edited by S.-K. Kim

101 **Shoulder and elbow trauma and its complications: Volume 1: The Shoulder**
Edited by R.M. Greiwe

Part One

Introduction to marine biomaterials

Discovery and development of marine biomaterials

1

B. Ben-Nissan

1.1 Introduction

Marine organisms are structured and constituted by materials with a vast range of properties and characteristics that may justify their potential application within the biomedical field. Moreover, assuring the sustainable exploitation of natural marine resources constitutes a highly interesting platform for development of novel biomaterials, with both economic and environmental benefits. In this perspective, an increasing number of different types of compounds are being isolated from aquatic organisms and transformed into profitable products for health applications, including controlled drug delivery and tissue engineering devices. A vital part of making regenerative medicine more of a future clinical success is the production of highly proficient scaffolds that function at the nanoscale, microscopic, and macroscopic levels. They structure and pull together cells into tissues, release encapsulated chemical signals in a targeted way and convey them into the body. There are abundant sources of structures and materials that can be used for a different function to their evolved intended one. The simplest strategy is to select a predesigned, preformed structure such as unique marine structures but modify it in a directed way specifically for its new intended function (Ben-Nissan and Green, 2013). Additionally, we can study nature and try and faithfully copy the vital components and reinvent this in the laboratory. In addition, we endeavor to learn more from nature the principle of low energy usage in the construction process, importance of structural organization, and implementation of transformative self-assembly and nonequilibrium chemistry.

These materials and designs have been instrumental to introduce the simplest remedies to vital problems in regenerative medicine, providing frameworks and highly accessible sources of osteopromotive analogues, nanofibers, micro and macrospheres and mineralizing proteins. This is exemplified by the biological effectiveness of marine structures such as corals and shells and sponge skeletons to house self-sustaining musculoskeletal tissues and for the promotion of bone formation by extracts of spongin and nacre seashells.

Molecules pivotal to the regulation and guidance of bone morphogenesis and particularly the events in mineral metabolism and deposition similarly exist in the earliest marine organisms because they represent the first molecular components established for calcification, morphogenesis, and wound healing. It emerges that bone morphogenetic protein (BMP) molecules – the main cluster of bone growth factors for human bone morphogenesis – are secreted by endodermal cells into the developing skeleton. Signaling proteins, TGF and Wnt-prime targets in bone therapeutics, are present in

Functional Marine Biomaterials. http://dx.doi.org/10.1016/B978-1-78242-086-6.00001-7
Copyright © 2015 Elsevier Ltd. All rights reserved.

early marine sponge development. Furthermore, ready-made organic and inorganic marine skeletons possess a habitat suitable for proliferating added mesenchymal stem cell populations and promoting clinically acceptable bone formation.

This chapter reviews the nature, morphology, and extent of this association and use of these structures for bone grafts, drug delivery, and extracts such as proteins for regenerative medicine. The role model for this specific biomimicry is the filtering microskeleton of Foraminifera. New selected strategies based on our and others' works for the engineering of new bone based on biomimicry will be also covered.

1.2 Marine environment

The marine environment and nature offers a number of potential intricate solutions to engineering and medical technologies at least in two simple basic ways: learning from nature allows us to emulate designs that are more efficient and functional and they are the resource of both inorganic and organic materials with a range of properties. Their unique assemblies can help us to understand mechanisms and reutilize the ideas and concepts for future applications. Nature is never perfect but its products function in a very efficient way. Trawling through natural materials has led to the discovery of ones with unique and unusual properties that can be directly exploited for technological gain.

During the last two decades, the main emphasis on marine-based structures has been on the hard and soft tissue engineering, efficient designs, property improvements, new pharmaceuticals, drug delivery and on discovery of new generation organic molecules. Proteins and biopolymers produced by marine organisms have been increasingly investigated for the biomedical applications. Natural history collections, specifically marine structures, are a unique and rich source of practical ideas and solutions for the initial stages of tissue reassembly in artificial culture.

Natural historians have compiled a record of structures, processes, and operations from every living organism. The translation of products from nature into technology is fundamental and the most powerful and successful way of resolving technological and scientific problems and conflicts.

1.2.1 Evolution

Evolution by natural selection has consistently originated assembly rules and design solutions that have been conserved and reapplied in organisms throughout the gradual ascendency and emergence of new forms of life (Ingber, 1998).

Natural history is the systematic study of nature and natural phenomenon using detailed observations. We can study evolution by natural selection through natural history or through embryology. Studying evolution of tissues and organs and their function and design is not a fully explored route to providing elements that can be used to reconstruct tissues in the simplest and most efficient way possible.

Nature is a consummate problem solver. We see exquisite, what seem to us as perfect, designs all around us. There are now systematic ways of harnessing these designs for a range of challenges in engineering and in clinical medicine. While nature cannot produce the *perfect* designs, it can generate the most ideal, optimized and functional adaptive ones.

Biological structures and biomaterials have evolved by natural selection over many millions of years of strict conditioning, tradeoffs between conflicting demands and limitations of an environment in order to maximize efficiency. These conflicts are omnipresent but can be partially resolved to generate extremely well-functioning materials. The end results are designs with compromises that exhibit high levels of performance made with minimum use of energy.

Evolution by natural selection has experimented with hundreds of thousands, if not millions, of inventive solutions some good some bad (Parker and Townley, 2007). We can only perform a miniscule amount of experiments in an average lifetime. Study of evolution by natural selection has already been used to understand and explain health and diseases by describing why human biology is the way it is (Williams and Nesse, 1991). A Darwinian or evolutionary approach to regenerative medicine may provide a wealth of information on the design of new replacement tissue. The evolution of tissues by natural selection provides us with a view of how different strategies of development have been harnessed by organisms according to function. As a result, we should be able to provide simplified assembly strategies to recreate functional approximations of every human tissue. Alternatively, we can use computer programming to develop structures using fundamental building plans and algorithms.

An ideal approach would be to use programs that evolve according to the principles of evolution by natural selection in which mutation and selection would act on rules of how to plan a body (Dawkins, 1989). In that way, we can potentially create new structures parallel to natural ones. However, the most immediate and accessible utility of computer programming to the design and fabrication of scaffolds is by direct modeling of human cells and tissues in health and disease. Tissue engineers are programming computer models of cell-to-tissue transformation and all the functions of tissue development and invasion into scaffolds and using the information to better optimize the structure (Sengers et al., 2007).

An understanding of the interplay between all the elements that go into the regeneration of tissues can be theoretically modeled. According to Sengers et al. (2007), these elements include nutrient transport and usage, matrix synthesis, cell population changes, attachment, migration, cell-to-cell interaction, and cell-to-matrix interaction (Sengers et al., 2007). The most common techniques currently used are computational topology design and solid free-form fabrication to generate scaffolds with precisely defined architecture. While at the same time, structures are being designed using computer algorithms that select the most ideal arrangement of pores, channels, and struts for their mechanical fitness.

1.2.2 Biomimicry

In nature, biomaterials possess enviable properties such as complexity, sophistication, and miniaturization that are not (as yet) possible to fabricate in the laboratory (Sanchez et al., 2007; Green and Ben-Nissan, 2008). However, we are gradually inventing ways of replicating nature to produce similar levels of sophistication albeit to a limited extent. Current 3D printing methods are good examples but we are only able to recreate microscopic structures with some level of biomimetic detail. This has been particularly true for the replication of bioinorganic structures. One versatile approach has been to use biological microstructures as templates for the reproduction of inorganic structures with identical features. In this sense, bacterial filaments have been used to make ordered silica microstructures and the tobacco mosaic virus to make nanotubes (Mann, 1983). As suggested by Mann, the main aim in biomimetics is to synthetically replicate the structures of selected inorganic biomatrices. Although not exclusive, they have clear role in the production of replacements for calcified tissues. This is achieved by using techniques in biomineral-inspired materials chemistry. The concept is to exploit the consecutive developmental pathway of systems that nature employs to make skeletons from molecules into macroscopic structures. Foundations are laid which define the space for construction. All the necessary building materials are supplied in a continuous delivery and keep pace with construction. The process in nature begins with supramolecular preorganization, interfacial recognition, vectorial regulation lending to multilevel processing (Green and Ben-Nissan, 2010). These processes are developed within confined reaction spaces directed in their formation by the templates themselves. The continual multiplication of these assemblies builds up into the emergence of morphology and macroscale biomimetic forms. In the construction of the simplest skeletons, a number of investigators have made great use of emulsion droplets to generate porous hollow shells, foams, and bead templates alongside the use of biocontinuous microemulsions to generate microskeletal networks (Gonzalez-McQuire et al., 2005).

Another strategy employs the controlled mineralization of adapted organic matrices from natural skeletons (Mann, 2001). These bioreplication strategies and approaches are being pursued with cellular and ECM inputs and include mineralization of reverse microemulsions (Green et al., 2004), mineralization of biliquid foams, mineralization of bicontinuous microemulsions (Walsh et al., 2005; Walsh and Mann, 1996), and template-mediated biomineralization of organic biomatrices (Hall et al., 2006).

This approach has yielded clinically relevant end results. Biomimetic microspheres created within self-organizing microemulsions were routinely synthesized and served as highly functional constructs for the localized delivery of growth factors and genes to primary human cells (Green, 2004; Figure 1.1). These unique particles were also proficient at producing osteoid and neocartilage.

1.3 Growing biomaterials with living cells

We are at the beginning of new biomimetic material synthesis systems with living cells with the obvious genetic instruction to direct the construction of biological materials and structures and tailor their intended function. Therefore, one of the most fascinating bioinspired approaches is to directly use cells and organisms to grow biomaterials and grow them to our specifications and requirements literally in the beaker or test tube (Parker and Townley, 2007). This can be achieved by judicious modulation of the growing environment. Single-celled organisms such as diatoms, *Foraminifera*, and *Coccolithophores* (Figure 1.2) are a convenient starting point as they are the most

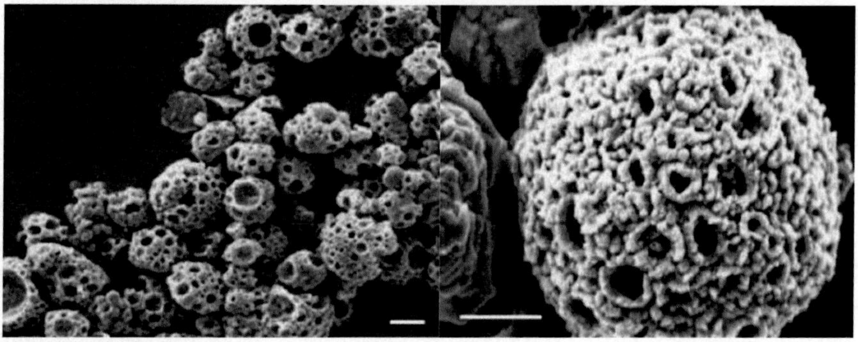

Figure 1.1 Biomimetic calcium carbonate microspheres made by bioinspired materials chemistry (scale 10 μm).

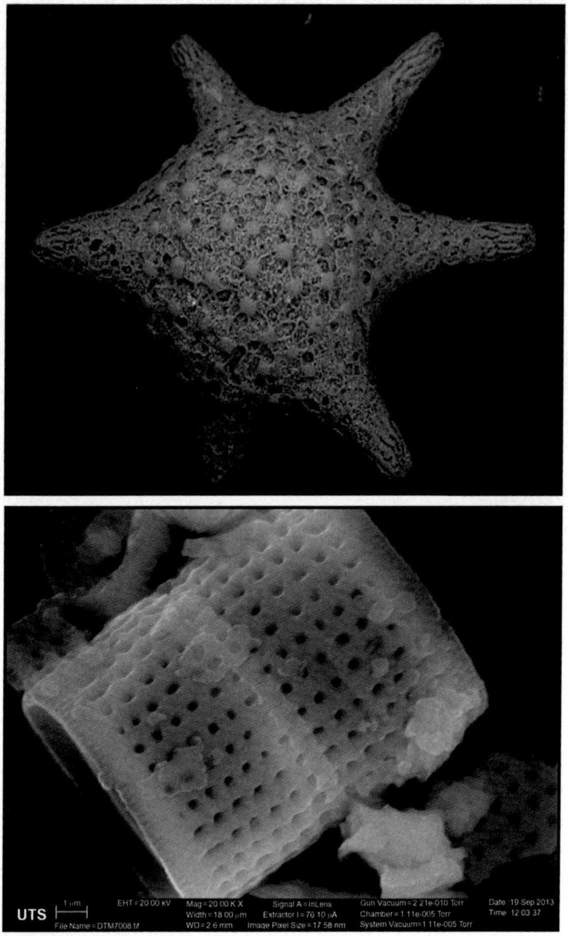

Figure 1.2 Foraminifera (top) and diatom (bottom).

rudimentary and elementary organisms to grow and support in artificial culture and provide enough utility for proving this approach as practically beneficial. Diatoms are of great interest for the development of new strategies in nanotechnology and molecular assembly as they provide modes of construction at these scales that could benefit the development of lab-on-chip biosensors, new generation drug delivery devices and bioreactors.

Diatoms are photosynthetic secondary endosymbionts found throughout marine and freshwater environments and are believed to be responsible for around one-fifth of the primary productivity on Earth. The genome sequence of the marine-centric diatom *Thalassiosira pseudonana* was recently reported, revealing a wealth of information about diatom biology. It has been suggested that diatoms can be used as drug-eluting modules because of their beneficial microscopic size and reticulated internal pore structure. In addition, the diatoms inspire the fabrication of nanostructured templates for nanoimprint processes (NIL), where large structured areas with nanometer precision are required. Belegratis et al. exploited the technical capabilities of the precise 3D laser lithography (3DLL) based on two-photon polymerization of organic materials. This method enabled the fabrication of arbitrary artificial diatom-inspired micro and nanostructures and the design of an inverse structure. Therefore, only one replication step is needed to obtain a template for NIL (Belegratis et al., 2014).

In the field of biomimetic photonic materials, there is a vision to grow materials with living cells integrated during synthesis and construction. This is an attractive proposition. In this way, the directed evolution may be possible with specific organisms that rapidly reproduce so that many thousands of generations are produced in short experimental periods. Protocols are well established now for the mass production of new proteins using a combination of site random mutagenesis followed by high-throughput screening (Kim, 2008; Green et al, 2013). Seemingly one of the most vital ideas about how nature constructs itself into the richness of shapes, sizes, and forms is called tensegrity (Ingber, 1998). In the vast diversity of nature, there are countless identifiable "ground" plans about how to construct and organize cells and tissues into organs.

1.3.1 Embryology

The spatial and temporal orchestration of cells to form tissues and organs is genetically controlled during embryology (Marga et al., 2007). Using the principles of developmental biology can inform new methods in tissue engineering. One example, as proposed by Marga et al., is a scaffold-free approach that relies on directing cells and tissues to self-organize and self-assemble. Embryological development teaches us how cells associate with each other and the effect of biochemical components.

1.4 Tissue engineering

Tissue engineering, and specifically synthetic tissue biology, is a newly emerging discipline which seeks to engineer tissues and form them into complex biological assemblages (Ben-Nissan, 2004). One approach in this method is to reverse-engineer natural biological materials, tissues, organs, or systems to "decipher" how they are put together and how they operate at highest level of detail. Some suggest that it will revolutionize the concepts and approach for re-engineering biological systems. Synthetic biology for forming multicellular tissues uses the most advanced methods available for building extracellular environments to direct morphogenesis of cells and tissues. In another way, cells are designed and constructed with novel functions and coaxed into multicellular organizations. Marine structures, as model templates, fit well to this new approach. There is also an aim to extend and modify the behavior of organisms (primarily unicellular ones because they are open to manipulation) and make them assemble and perform new programmed tasks, some natural and others unnatural functions (Andrianantoandro et al., 2006).

1.4.1 Natural skeletons

Natural and calcium phosphate-converted corals have been the primary source of natural skeletons for bone tissue engineering because of their chemical, crystallographic, and structural compatibility to native human bone (Ben-Nissan, 2003). Using natural skeletons in a direct way as a scaffold for growing cells into tissue emerged for making new bone tissue as a product of hydrothermal processing (Weber et al., 1969, 1971; White et al., 1972; Roy and Linnehan, 1974). For the same reasons, since then, researchers have made use of invertebrate marine skeletons of hydrozoans, cuttlefish (Rocha et al., 2005), marine sponges (Green et al., 2004), nacre seashell, and echinoderm spines (Martina et al., 2005) as templates with optimal ranges of pore sizes, channels, and structural networks for organizing and nourishing the growth of human tissues as a prelude to transplantation into the patient. In other developments, whole natural skeletons have been used as templates for carrying biomolecules. Accordingly, diatom skeletons have been tethered with active biomolecules, such as an antibody, to be used in immunodiagnostics (Townley et al., 2008).

1.4.2 Scaffolding development

In search of scaffolding materials, we have so far identified candidate biomatrices in nature, with varied chemical homologies and structural analogies to human extracellular matrices and whole tissues. They include nacre marine shell, marine sponge skeletons, echinoderm skeletal elements, and coral skeletons. The utility of selected species of these marine animals has been applied to the regeneration of human bone and cartilage. However, the full utility in these tissues and other tissues has yet to be harnessed and exploited.

The regeneration potential of human bone is limited in the case of repair of large bone defects. In most cases, autogenous and allogenic bones are used as bone grafts. However, the amount of autogenous bones that can be utilized within the body is limited and there is a risk of donor site morbidity. In addition, there is a possibility of transmission of infection from allografts if they are based on demineralized bones. Therefore, synthetic bone graft materials, such as calcium phosphate ceramics, polymers, and composites, have been widely developed as bone graft materials (LeGeros, 1993).

Hydroxyapatite (HAp) [$Ca_{10}(PO_4)_6(OH_2)$] is the main inorganic mineral constituent close to human bone chemistry, and is also an outstanding synthetic bone substitute because of its osteoconductive properties. HAp ceramics can be manufactured synthetically from its constituents via a range of production methods. In addition, they have been manufactured by demineralizing bovine or human hard tissues.

Calcium phosphates are also prepared from natural marine structures such as corals (Roy and Linnehan, 1974; Vago et al., 2002; Papacharalambous and Anastasoff, 1993), mussel (Macha et al., 2013), sea shells (Bahar et al., 2003), sea urchin (Vecchio et al., 2007; Samur et al., 2013), land snail shells (Kel et al., 2012), cuttlefish bone (Rocha et al., 2006), and pearl (Shen et al., 2006) to name just a few. HAp powders have commonly been prepared using a variety of techniques such as wet chemical synthesis, hydrothermal conversion, solid-state reaction, and calcination of bone.

Based on observed tissue response, synthetic bone-graft substitutes can be classified into inert (e.g., alumina, zirconia), bioactive (e.g., hydroxyapatite, bioactive glass), and resorbable substitutes (e.g., tricalcium phosphate, calcium sulfate). Of these, resorbable bone-graft substitutes are preferred for bone defect filling because they can be replaced by new natural bone after implantation. β-tricalcium phosphate ($Ca_3(PO_4)_2$, β-TCP) is one of the most widely used bone substitute material, due to its faster dissolution characteristics. Preparation of magnesium-substituted tricalcium phosphate ($(Ca, Mg)_3(PO_4)_2$, β-TCMP) has been reported by precipitation or hydrolysis method in solution. These results indicate that the presence of Mg stabilizes the β-TCP structure (LeGeros et al., 2004). The incorporation of Mg also increases the transition temperature from β-TCP to α-TCP and decreases the solubility of β-TCP (Elliott, 1994; Ando, 1958).

Spines of sea urchin *Heterocentrotus mammillatus* and *Heterocentrotus trigonarius*, mainly consist of large single crystals of Mg-rich calcite ($(Ca, Mg)CO_3$) (Vecchio et al., 2007; Weber et al., 1971; Su et al., 2000; Politi et al., 2004). These single crystals usually have smooth, continuously curved surfaces that form a three-dimensional fenestrated mineral network (Figure 1.3).

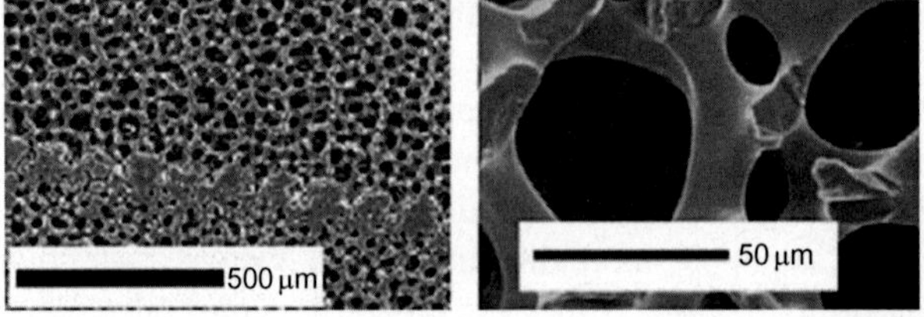

Figure 1.3 SEM images of spines of a sea urchin: a section normal to *c*-axis shown at the left, the portion at right shows the porous nature of the same figure at a larger magnification.

1.4.3 Sponge skeletons

Marine sponges possess the most primitive form of extant tissue (600 million years old), but share much in common with multicellular tissues, which have apparently conserved many features evolved by these first multicellular organisms. Morphological and biochemical similarities exist between marine sponge and vertebrate extra cellular matrix (ECM) alluding to fundamental rules of organization evolved first by marine sponges. Three collagen types have so far been identified from marine sponge. All sponges are composed of 22 nm thin collagen fibrils with highly ordered periodic banding. Although the collagen ultrastructure is relatively simple compared to vertebrate collagens, amino acid sequences and genome organization are similar. Collagen fibrils are secreted in bundles in a similar manner to vertebrates. Similarly collagen fibrils are closely associated with proteoglycans which, in mammalian tissue blueprint, shape, and form at long range scales. Fibronectin, dermatopontin, and tenascin polypeptides are also found in marine sponge collagen fibers and cross-react with antibodies raised against vertebrate analogues highlighting their common origins. Some sponge species possess an analogue of type IV collagen found in vertebrate basement membrane collagens (Boute et al., 1996). The organization of collagen fibrils is analogous to collagen type XIII, which sticks cells to surfaces (Exposito et al., 2002).

It is with these properties (fibronectin and cell adherent collagens) that collagenous marine sponges represent a significant potential for future development as bioactive tissue engineering scaffolds.

At present, marine sponges are extensively exploited for novel biological compounds as potential treatments for cancer tumors, leukemia, and inflammation. Marine sponges are also a source of collagen for cosmetics (Swatschek et al., 2002) and dermatological preparations (Nicklas et al., 2009).

In total, 50% of all marine-derived materials are sourced from a wide spectrum of marine sponges. Collagenous marine sponge skeletons are incredibly soft, strong, highly absorbent and elastic, resistant to high temperatures and bacterial attack. Such properties make them highly suited for surgical procedures. Tissue formation within an implanted collagenous marine sponge from embryonic cell is shown in Figure 1.4. The exact conditions to grow marine sponges at a large enough scale for commerce are being investigated by a number of researchers. Some have established aquatic pilot farms for the cultivation of selected bath sponge species. Marine sponges are sufficiently adaptable for commercial scale production. Another aim for cultivating marine sponges is to extract medically important secondary metabolites in much larger quantities than is possible from collections made by conventional bioprospecting. Marine sponges are also being investigated in cutting edge research for the design and fabrication of new advanced functional silica materials for use as more efficient conductors of light and therefore as models for future fiber optics. The superior optimized structural design of silica marine sponges (Figure 1.5) has been alluded to and which provides useful lessons for construction of man-made frameworks with minimal starting materials for maximum strength (Aizenberg and Hendler, 2004). There is therefore considerable mechanical benefit to specific arrangements of structural elements at many different hierarchies of scale.

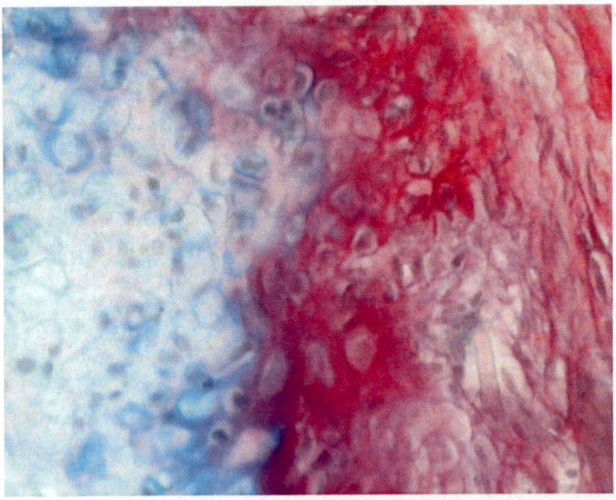

Figure 1.4 Tissue formation within an implanted collagenous marine sponge from embryonic cell.

Figure 1.5 Fine intricate structural design of silica marine sponges (Glass sponge).

1.5 Learning crystallography from sea urchin

The teeth and the bones of mammals, the protective shells of mollusks, and the needle-sharp spines of sea urchins (Figure 1.6) and other marine creatures are made from calcium carbonate (calcite or aragonite) and/or calcium phosphate-based materials.

The skeleton of sea urchin spines is composed of large single crystals of Mg-rich calcite, which have smooth, continuously curved surfaces and form a three-dimensional fenestrated mineral network. Spines of the echinoids *H. trigonarius* and *H. mammillatus*

Figure 1.6 The sea urchin spicules.

can be converted by the hydrothermal reaction at 180 °C to bioresorbable Mg-substituted tricalcium phosphate (β-TCMP). Due to the presence of Mg in the calcite lattice, conversion of β-TCMP occurs preferentially to hydroxyapatite formation. The converted β-TCMP still maintains the three-dimensional interconnected porous structures of the original spine. It is thought that the main conversion mechanism is the ion-exchange reaction, although there is also a dissolution–reprecipitation process that forms some calcium phosphate precipitates on the surfaces of the spine network. The average fracture strength of urchin spines and converted spines (β-TCMP) in the compression tests are 42 and 23 MPa, respectively. *In vivo* studies using a rat model demonstrated new bone growth upto and around the β-TCMP implants after implantation in rat femoral defects for 6 weeks. Some new bones were found to migrate through the spine structural pores, starting from the outside of the implant through the pores at the edge of the implants. These results indicate good bioactivity and osteoconductivity of the porous β-TCMP implants.

Sea urchin skeletal plates are punctured by a very regular series of pores. Approximately three quarters of the pores are exits for tube feet (200 μm pore diameters at the spine bases to 600 μm pore diameters for the tube feet in *Centrostephanus nitidus*) while the remainder are channels connected to the reproductive and alimentary systems and are very much larger to accommodate larger throughputs of fluid (1000–2000 μm pore diameter). Similar to teeth and bones, the sea urchin spicule is a composite of organic and inorganic materials that the animal synthesizes using the most readily available elements in sea water. The fully formed spicule is composed of a single crystal with an unusual morphology. It has no facets and forms a star-like shape. To achieve such unusual morphologies, sea urchin and other marine organisms deposit a disordered amorphous mineral phase first, and then slowly transform into a crystal with neatly aligned lattice with a specific and regular orientation, while maintaining their morphology (Politi et al., 2004).

A unique transformation from disordered amorphous structure to ordered crystalline structure occurs. These amorphous to crystalline phase transformation from the liquid state can teach us to produce highly ordered and well-oriented synthetic nanomaterials and composites.

The sea urchin spicule is formed inside a clump of specialized cells and begins as the animal lays down a single crystal of calcite, from which the rest of the spicule is formed. Starting from the crystalline center, three arms extend at 120° from each other. The three radii are initially 40–100 nm-sized amorphous calcium carbonates, but slowly convert to well-organized and -oriented calcite. Mechanism is not clearly understood but might be through dissolution and ordered precipitation mechanism at known crystallographic orientations of calcite or aragonite.

1.5.1 Amorphous hydroxyapatite to crystalline nanoplatelet formation

In order to simulate the amorphous to crystalline transformation, Milev et al. developed a novel method to produce single-phase, nano-sized, plate-like, mixed A–B-type carbonate-containing apatite (CAp) similar to bone apatite by using sol–gel technology (Milev et al., 2003). The methodology emulates biomineralization, where topotactic transition from amorphous to octacalcium phosphate (OCP) than to hydroxyapatite (HAp), which is believed to occur *in vivo*.

The synthetic process developed involves formation of thin (1–1.4 nm) layered calcium phosphonate salts by a self-assembly process. The phosphonate-derived apatite shows slightly curved plate-like amorphous apatite, which can convert to a crystalline form with significant changes to both (a) and (c) crystallographic axis as function of temperature (Figure 1.7). The thermal decomposition of these layered amorphous salts leads to formation of highly crystalline plate-like carbonated apatite (Milev et al., 2003). The overall carbonate content varies from 6.4 to 4 wt%, within the temperature range of 500–700 °C. This carbonate content corresponds well with the amount found in mammalian hard tissues. Although sea urchin transformation is carried out in normal sea water temperature ranges, this similar transition in hydroxyapatite shows a plate-like morphological adaptation without shape change while the amorphous to crystalline phase transformation occurs.

Raz et al. (2003) reported the transient phase of amorphous calcium carbonate in sea urchin larval spicules and the involvement of proteins and magnesium ions in its formation and stabilization.

This was further confirmed by a study of Gong et al. (2012) using X-ray absorption near-edge structure (XANES) spectroscopy and photoelectron emission microscopy

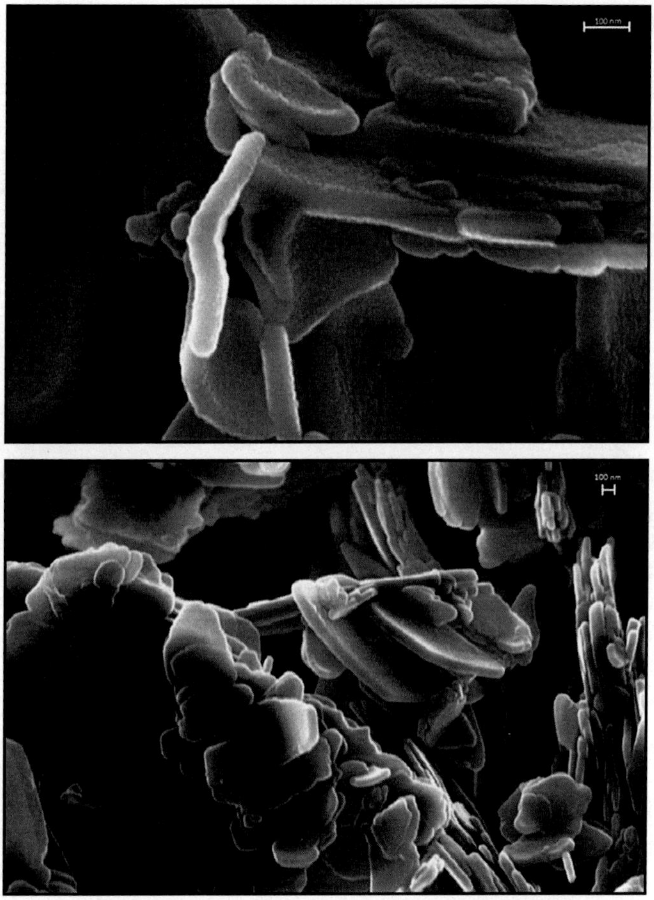

Figure 1.7 Sol Gel derived amorphous hydroxyapatite platelets (top figure) and their full conversion to crystalline hydroxyapatite (bottom figure) after an appropriate 300°C heat treatment.

(PEEM). They examined forming spicules in embryos of *Strongylocentrotus purpuratus* sea urchins, which showed that a sequence of three mineral phases occurred: hydrated amorphous calcium carbonate ($ACC \cdot H_2O$) → dehydrated amorphous calcium carbonate (ACC) → calcite transformation. Unexpectedly, they found $ACC \cdot H_2O$-rich nanoparticles that persist after the surrounding mineral has dehydrated and crystallized. Protein matrix components occluded within the mineral was thought to inhibit $ACC \cdot H_2O$ dehydration. They further devised an *in vitro* experiment, using XANES-PEEM, assay to identify spicule proteins that may play a role in stabilizing various mineral phases, and found that the most abundant occluded matrix protein in the sea urchin spicules, SM50, stabilizes $ACC \cdot H_2O$ *in vitro* (Gong et al., 2012).

1.6 Nacre

For the modern day clinical applications, the outer nacreous layer of a particular species of mollusc shell is an unlikely and unexpected source of biomaterial for engineering new bone.

The scientific basis of fusion with bone was first uncovered by Lopez et al. (1992) and Lamghari et al. Under closer scrutiny, nacre was found to activate skeletal cells, induce bone formation, and provide structural support in a human clinical trial (Lamghari et al., 2001).

Unlike any other biomaterial, nacre from the pearl oyster, *Pinctada maxima*, is able to induce osteogenesis and bone formation from latent osteoprogenitors along an endochondral pathway, consisting of a cartilage tissue intermediary phase (Lopez et al., 1992).

The outer nacreous layer of a particular species of mollusc shell is an unlikely and unexpected source of biomaterial for engineering new bone.

Natural biomaterials were widely used by physicians from ancient civilizations of India, China, Egypt, and Central America and it was the ancient Mayans who discovered the unique property of nacre to heal seamlessly onto living human bone without causing harm.

Nacre has been tested in human, sheep, and rabbit models (Lamghari et al., 2001). In human patients, fresh woven bone bonds itself throughout the nacre implant, augmented by the heightened activities of osteoblasts and osteoclasts (Figure 1.8). While nacre is stably tolerated *in vivo*, its degradation and resorption is limited and this

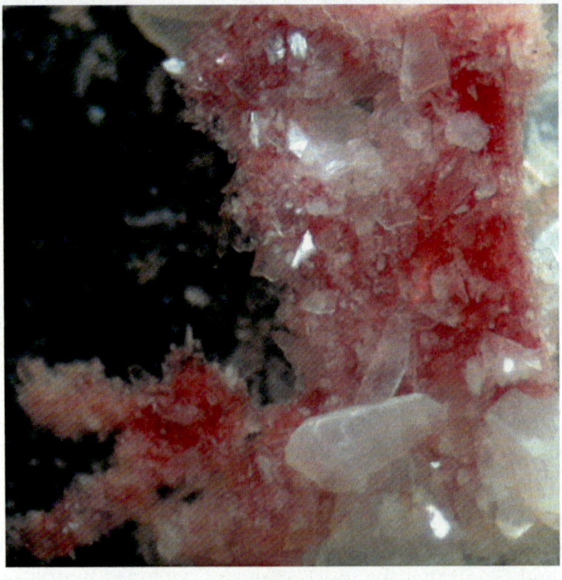

Figure 1.8 Combined *in vitro* culture of hBMSC. The cell mass is stained red for ALP secretion (a primary marker of bone formation). Nacre chips of various sizes can clearly be seen.

could hinder its use within calcified tissue requiring rapid self-renewal (Lamghari et al., 2001). Although somehow controversial in definition according to nacre researchers, the "water-soluble matrix fraction" (WSM) of nacre directly induces bone formation (Rousseau et al., 2003). Molecules from nacre matrix have been shown to decrease bone resorption by acting on osteoclast metabolism (Duplat et al., 2007). The available evidence suggests that mobile signal transmitters involved in the biological control of mineralization (as an initiator and inhibitor of calcium carbonate crystallization at the mineralizing growing front) dissolve into solution-induced differentiation of surrounding latent osteoprogenitor cells (Westbroek and Marin, 1998). The reason why nacre directly induces human cells to form new bone is best explained by the idea that "a signaling" biomolecule is involved in regulating cell-mediated biomineralization is common to both vertebrate bone tissue and nacre. These biomolecules must have been therefore conserved by evolutionary selection pressures.

Mollusc shells are a fascinating model for understanding the complexities of biomineralization such as the control and regulation of protein–mineral interactions. Knowing the proteins involved is a prerequisite for designing advanced functional materials. So-called osteopromotive effect, as measured by ALP expression, of nacre is also commensurate with treatment with dexamethasone, at least in fibroblasts. Size exclusion HPLC of the water-soluble matrix has uncovered protein fractions rich in glycine and alanine, with specific biochemical effects on human fibroblasts that modulate cell differentiation and proliferation (Almeida et al., 2001). Peptides are prevalent in the nacre matrix. Particular individual fractions have been shown to give rise to specific responses from cultured osteoblast cells. Protein fractions with low-molecular weight (less than 1 kDa) for example, up-regulated ALP secretion whereas high molecular weight fraction reduced ALP secretion. Detailed sequencing of water-soluble proteins using proteomics offer enhanced characterization of nacre matrix proteins. Nacre WSM was also shown to increase the secretion of a key inhibitor of apoptosis, cytoplasmic Bcl-2 and has an influence on rat calvarial osteoblast maintenance and survival. Low-molecular weight (LMW) fractions were recently found to increase expression of collagen type I and the osteogenic-associated mRNA expression of osteopontin and Runx-2 (Rousseau et al., 2007).

Further detailed characterization of the bioactive LMW molecules has led to the identification of 110 molecules in the 100–70 Da range comprising of glycine-enriched peptides with structural similarities and high affinities for each other. A highly defined matrix protein with a 10-kDa size named as p10 has specifically demonstrated an increase in human fibroblast cell ALP expression (Zhang et al., 2006) lending greater hope that the osteogenic signal molecules can be isolated in their vital functional form. A soluble p60 protein conglomerate extracted from decalcified nacre possesses sufficient bioactivity on 3T3 and MSC to induce the secretion of mineral nodules. Some of the specific biomolecular mechanisms and associations between the signal molecules and cellular processes are been gradually uncovered (Green et al., 2013).

Some biomineralization researchers are doubtful that nacre proteins are the primary cause of osteoinduction. In a study by Liao et al. (2002), nacre failed to stimulate an *in vivo* osteogenic response, although bone-to-nacre apposition and bonding did occur directly. Liao et al. (2002) suggested that nacre provided a favorable surface chemistry, rich in phosphorous, favorable to osteoclast and osteoblast recruitment, attachment, and matrix synthesis. In an *in vivo* ectopic bone environment, surface-modified nacre was found not to be osteoinductive within demineralized bone matrix, but its integration and fusion with bone was better than non-nacre controls. In another study, Kim et al. (2000) investigated the role of interfacial properties on the biocompatibility of nacre and specifically its unique bone-bonding ability. Kim and colleagues concluded that the organic matrix is what makes nacre bond to bone so well, as it creates a favorable surface charge for optimal biological associations. When implanted, the organic matrix of nacre is thought to generate a new interfacial microenvironment that forms many functional associations with the surrounding tissue leading to a better bone bonding than bioceramic implants without an organic matrix. The WSM was implicated in the formation of this HAp layer and the augmented cell responses (Rousseau et al., 2003). Taken altogether, nacre provides an appropriate tissue-compatible physical platform, which shows unique peptides that initiate and drive bone formation.

In addition, nacre due to its organic content and plate-like design is mechanically tough (fracture toughness equivalent to titanium), nonimmunogenic, and rapidly biodegradable, without eliciting detrimental physiological effects. These characteristics of nacre offer us a unique substrate for delivery of a functional (possibly osteopromotive) agent to sites of bone loss in quantities that lead to rapid bone repair and regeneration.

1.7 Echinoderm skeletal elements

In the end of 1980s, a number of studies using the replamineform technique for replicating perforate reef-building coral skeletons and echinoderm structural elements to generate hard tissue replacements to bone, as well as candidate prostheses for blood vessels and trachea were introduced (White et al., 1972). In this context, the skeletal ossicles from the sea star (*Pisaster giganteus*) have been investigated. They provide an ideal architecture together with physical and chemical properties conducive to bone restoration (Martina et al., 2005).

Echinoderm skeletons are constructed from a unique, intricately shaped, 3D, single crystalline meshwork with a topological structure in which every internal pore and channel is in direct contact with all others (periodic minimal surface). This property is likely to facilitate mass transfer and tissue development (Aizenberg and Hendler, 2004).

Hydrothermal or similar conversion of calcium carbonate to calcium phosphate processing of echinoderm structures as with other mineralized skeletons such as corals, transforms the chemical and mechanical properties with equivalence to human bone (Hu et al., 2001; Vago, 2008).

Studies are of their way to learn and adapt from the natural materials the ways they adapt to their environment which can enable us to produce an exciting array of self-responsive structures and materials for regenerative medicine. In nature, biomaterials are made with immaculate resource and energy efficiency using common, readily available substrates through self-assembly into highly organized hierarchies.

Functional natural structures optimized to their environment are produced by self-assembly. This gives us the opportunity to produce highly efficient structures that can minimize energy use, with intricate shapes and architectures that are tailored to their functions and do not break down.

Biomimetic approach has been very successful in showing that they can yield promising outcomes for application in the tissue engineering of skeletal tissues. One such approach involves configuring material environments at the molecular and macromolecular scales that try to mimic native extracellular matrix. The hope is to further extend this ongoing research toward the design of clinically relevant scaffolds for regenerative medicine using a unique set of "self-organizing hierarchical structures" designed and synthesized according to biological principles. In this instance, we begin by harnessing as nature does, with enormous precision and control of inorganic molecules, nanofibres, and nanoparticles to construct advanced functional biomaterials with application in tissue engineering and targeted pharmaceutical drug and gene delivery for a range of clinical applications.

1.7.1 Coral skeletons

Natural coral exoskeletons have been used widely as a bone replacement in orthopedic, craniofacial, dental and neurosurgery owing to their combination of good mechanical properties of open porosity (Figure 1.9) and ability to form chemical bonds with bone and soft tissues *in vivo* (Ben-Nissan, 2003). In general, corals have the best mechanical properties of the porous calcium-based ceramics. According to Dauphin, the organic composition has an important part to play in coral biocompatibility. The abundance, conformation, and composition of the organic matrices are responsible for successful biological integration of coral with human host (Vago, 2008).

The beginning of the coral life cycle starts with the polyps, which absorb the calcium ions and carbonic acid present in the seawater to produce the calcium carbonate in the form of aragonite crystals. The remaining composition consists of trace

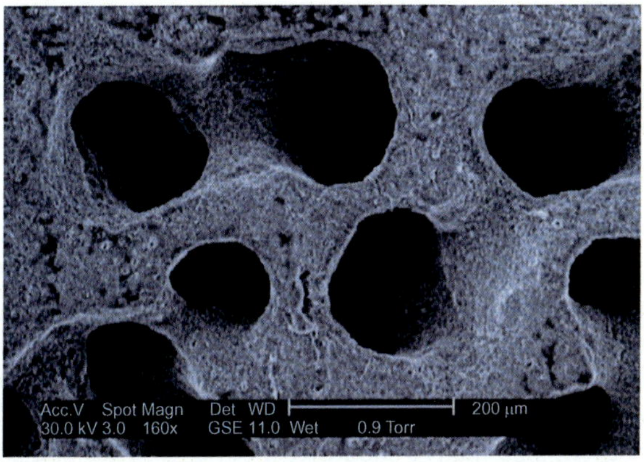

Figure 1.9 Coralline apatite converted by hydrothermal process showing interconnected porous structure.

elements of magnesium, strontium, fluorine, and phosphorous in the phosphate form (Stanley, 2003). Once implanted in the human body, these elements play a critical role in the bone mineralization process and in the activation of key enzymes associated with bone remodeling cells. Strontium has been shown to contribute to the mineralization process by stimulating osteoblasts while inhibiting osteoclasts (Bonnelye et al., 2008). Similarly, fluorine helps bone formation through similar stimulatory effect on osteoblast proliferation. Magnesium is also long known to be beneficial in bone remodeling as it has been shown to increase the mechanical properties of newly formed bone (LeGeros, 1981).

The organic composition has an important part to play in coral biocompatibility. The abundance, conformation, and composition of the organic matrices are responsible for successful biological integration of natural coral with human host (Green et al., 2013).

Use of coral skeletons for general routine orthopedic surgery and tissue engineering has been so far limited to external fixation devices as they are inappropriate for strictly load-bearing applications die to their calcium carbonate structure with high dissolution rates. Sol–gel-coating technologies can be used to enhance the strength of corals and this enables them to be used at more skeletal locations (Ben-Nissan, 2004; Choi and Ben-Nissan, 2007).

Corals offer great opportunities to tissue engineering of bone either in their natural form or as hybridized synthetic forms. Coral skeleton combined with *in vitro* expanded hBMSC increased osteogenesis more than obtained with scaffold alone or scaffold with fresh marrow (Green and Ben-Nissan, 2010). *In vivo* large animal segmental defect both orthopedic and maxillofacial surgery led to complete recorticalization and formation of medullary canal with mature lamellar cortical bone and onlay graft for contour augmentation of the face giving rise to clinical union in a high number of cases (Papacharalambous and Anastasoff, 1993; Leupold et al., 2006).

Corals offer great opportunities to tissue engineering of bone either in their natural form or as hybridized synthetic forms. Coral skeleton combined with *in vitro* expanded hBMSC increased osteogenesis more than obtained with scaffold alone or scaffold with fresh marrow. *In vivo* large animal segmental defect studies led to complete recorticalization and formation of a medullary canal with mature lamellar cortical bone giving rise to clinical union in a high number of cases (Petite et al., 1996).

Structural and biomineralization studies of coral can be used to inform the development of new advanced functional materials because of the unique nanoscale organization of organic tissue and mineral as highlighted by Ehrlich et al. (2006). At a macrostructural level, the deep-sea bamboo coral exhibited bone-like biochemical and mechanical properties (Nowak et al., 2005). A specialized collagen matrix (acidic fibrillar) serves as a model for future potential tissue engineering applications. Figure 1.10 shows a widespread colonization of coral skeleton by primary mesenchymal stem cells stained green to show that they are viable and metabolically active. Opportunities in tissue engineering using coral skeletons have yet to be fully realized and exploited. The matrix supported both osteoblast and osteoclast growth and the exceptional bioelastomeric properties of the collagen matrix (gorgonin) of this coral make it potentially suitable for blood vessel implants. Quinones crosslink and

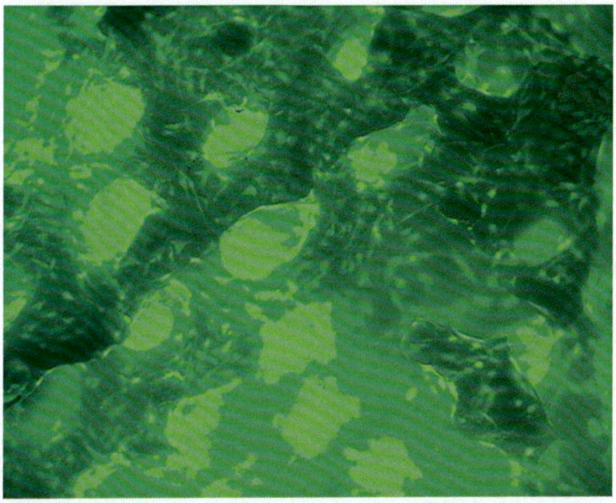

Figure 1.10 Showing widespread colonization of coral skeleton by primary mesenchymal stem cells stained green to show that they are viable and metabolically active.

harden the collagenous gorgonin proteins and closely resemble human keratin. The mechanism by which gorgonin is synthesized and interacts with the process of mineralization may provide lessons for the generation of a synthetic collagen-like material (Ehrlich et al., 2006).

1.8 Drug delivery and marine structures

Slow or targeted drug delivery system is a system that is capable of releasing a preloaded pharmaceutical agent to a targeted site at a specific rate and most importantly, at a therapeutically relevant concentration. The main aim of this type of system compared with conventional drug intake (injection or tablet) is to facilitate the local and specific area delivery, dosage, and duration control and hence appropriate active drug delivery while causing minimal side effects. While technological advancement has produced innovative and refined drug delivery systems, the fundamental basis that defines what a drug delivery system is remains unchanged. The therapeutic advantages of these systems can be attributed to many underlining factors: predictability of release rate, minimized drug concentration, thereby reducing any possible adverse systemic effect.

Prolonged duration of drug therapy such as the need for frequent redosing has been problematic in many global applications of drugs such as the treatment protocol of malaria in Africa. Many factors are considered in the development of drug delivery systems in accordance with the desired application. This includes the agent to be carried, the administration route, the material used, the degradation rate, the loading efficiency, the physical and chemical properties of the material, the practicality for large-scale production, toxicity, among other parameters.

Many materials such as ceramics, polymers, alginate, and polysaccharides have shown potential advantages as drug delivery systems. However, marine materials such as coral exoskeletons and marine shells show a better promise due to their easy conversion to fully resorbable calcium phosphates, intricate interconnected pores to load drugs and their controllable dissolution rates.

The pore size and interconnectivity of the coral pores are a critical factor in the rate of coral as a bone graft and slow drug delivery material. Moreover, the uniform porosity of the exoskeletons provides a more constant drug loading and therefore provides a more predictable drug release rate of which both are crucial factors that directly impact the effectiveness of the drug delivery system. Natural structures, although not perfect, often exhibit intricate morphologies that justify the efforts for biomimetic approach.

1.8.1 Foraminifera: template-mediated delivery

One significant worked example of biomimetic materials chemistry applied to the drug/gene delivery uses template-mediated mineralization chemistry within a complex organized 3D reaction field with patterning that mimic plankton shells. This chemical system is designed to integrate processes of "self-assembly and self-organization" in space. Constructs that are generated in this manner provide many distinct advantages for tissue engineering as a physical template and devices for controlled release of range of drugs, BMP, proteins, genes, and growth factors. Since the chemistry occurs at room temperature and consists of an aqueous phase, biological molecules can be safely incorporated during synthesis.

Another functional success has been the potential for drug incorporation and delivery from nano, micro, and macrospheres. During an overseas summer break, I have noticed that the sand that I was walking on the beach looked as perfect spheres and showed quite unique intricate structures (Figure 1.11). I brought some to our laboratories and after SEM and XRD analysis I noticed that it was not silica sand but calcium carbonate marine shells.

After proper characterization, we identified it as a marine structure belonging to the *foraminifera* family. *Foraminifera* are abundant and are found in all marine environments, but different species exist with different shapes depending on their environment. *Foraminifera* are single-celled organisms with shells consisting of multilayer inner chambers commonly divided and added during its growth.

These observations opened a new thinking in drug delivery. Natural spheres loaded with drugs can spontaneously degrade and progressively release entrapped biological contents introduced during synthesis.

We further collected spherical fossilized shells Floresianus (*Foraminafera*) from coral beach sand of Great Barrier Reef, Australia. The samples were intact, lacked spines, and measured 0.5–1.5 mm in diameter. These "shells" or more appropriately microspheres, possess unique fenestrated structures that have evolved to circulate seawater and collect light for the mutual benefit of symbiotic algal cells that reside inside the shell. Microcomputed tomography (μ-CT) and scanning electron microscopy (SEM) imaging confirmed that these shells were internally permeated by a 3D network of microscopic interconnected channels measuring 1–10 μm in diameter. Between the micropores, surface area included nano and mesopores.

Prior to any marine material can be used as a graft or drug delivery material, it must first undergo a rigorous process to test the composition, purity, morphology, and

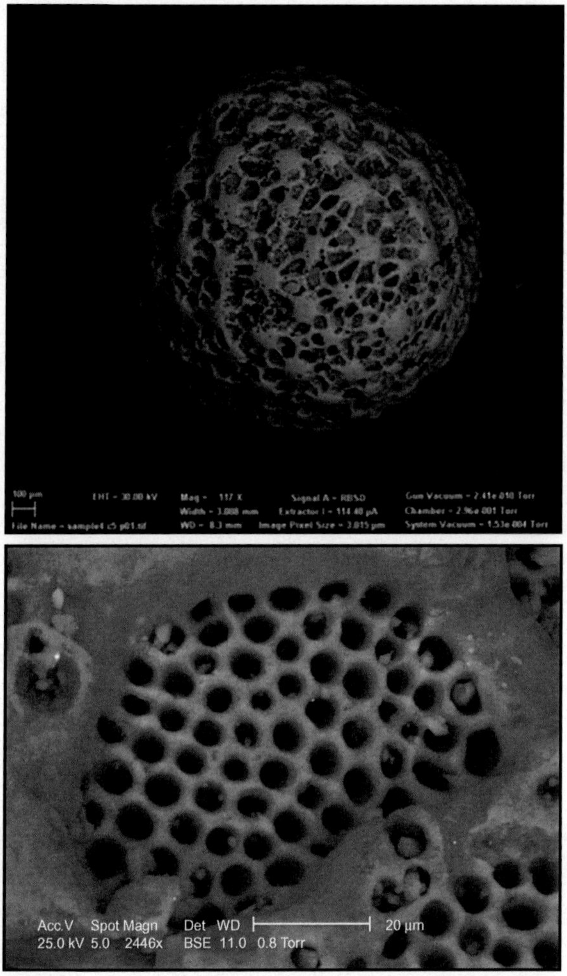

Figure 1.11 Foraminifera shells from Great Barrier Reef, Australia (top figure) and an enlarged surface micrograph (bottom) showing unique interconnected porosity structure.

suitability for drug loading and its slow dissolution without any adverse effect to the patient. Unless specifically protein and organic matter are required to be present, prior to sterilization of calcium carbonate material, any residual organic constituents are removed by immersing in solution of sodium hypochlorite then drying at about 100 °C. These shells can be hydrothermally converted into more stable, highly crystalline β-tri calcium phosphate (β-TCP) and hydroxyapatite (HAp) using the methods developed earlier (Hu et al., 2001).

Chemical conversion of these shells to calcium phosphate (microspheres) does not change the original structure making it available for adsorption of candidate drug compounds and which can allow new bone cell penetration into the micropores after bone graft implantation. Due to the pore architecture, these microspheres within the physiological environment can dissolve and supply calcium and phosphate ions and other ions and the drugs are incorporated to the immediate bone structure.

These and other converted natural spheres loaded with drugs can spontaneously degrade within the physiologic environment and progressively release entrapped biological contents introduced during synthesis. The release profile shows relatively slow, local release of drugs such as bisphosphonate, gentamicin, and simvastatin from micro and macrospheres for extended periods (Chou et al., 2011, 2012, 2013).

1.9 Marine structures and stem cell regulation

There is also growing realization that the composition of a scaffold material is vitally important to organize stem cell activities as they are dependent on the extracellular fabric for life support and to guide their subsequent evolution and development (Green et al., 2012; Abramovitch-Gottlib et al., 2006). This can be achieved by using natural structural biomaterials, their derivatives and reconstituted forms. Important aspects of regenerative medicine of significance are to improve stem cell processing, provide microenvironments that are better at regulating tissue formation and development and manufacture less invasive transplantation modules with site-specific targeting properties.

These can be demonstrated with recreating a mesenchymal stem/progenitor cell niche for regulating cell fate and proliferation. Figure 1.12 shows a foraminifera microsphere coated with rat adipose mesenchymal stem cells after 10 days of pellet culture. Figure 1.13 demonstrates mesenchymal stem cells spanning pores of a sea urchin spine.

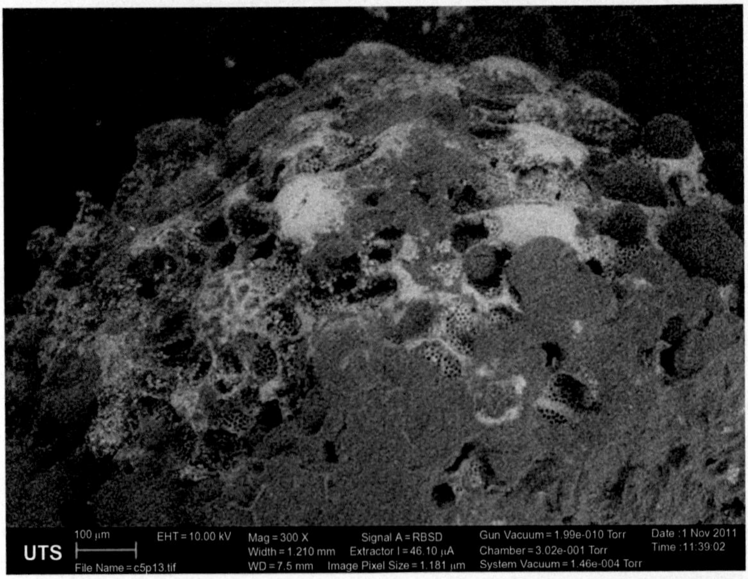

Figure 1.12 Rat adipose mesenchymal stem cells on foraminifera after 10 days of pellet culture.

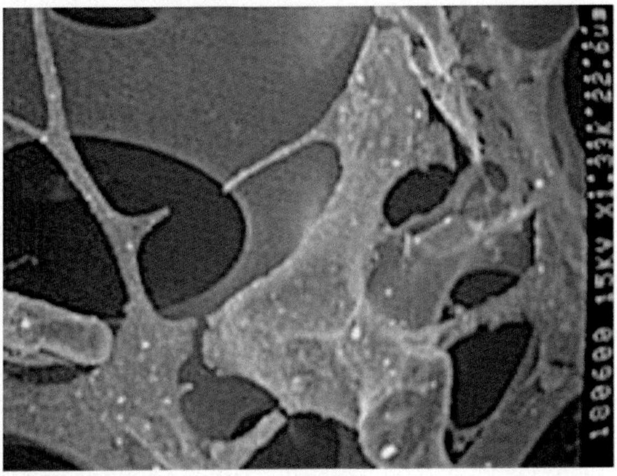

Figure 1.13 Mesenchymal stem cells spanning pores of a sea urchin spine.

To fabricate extracellular matrix equivalents with soft sugar (polysaccharide) hydrogels interpenetrated with networks of native extracellular matrix fragments and the synthesis of transplantation modules for higher resolution targeting is now a reality.

Recreation of the native stem cell environment where stem cells normally reside (such as bone marrow) are protected, managed, and stabilized as self-renewing undifferentiated cells and given instruction on how to regulate the rate of progenitor and successor cell production is an active research area. According to Nurcombe and Cool, the glycosaminoglycan heparin sulfate is the master molecule controlling the dynamics of almost all stem cell functions (Nurcombe and Cool, 2007; Grunert et al., 2008).

Making unique customized environment that can regulate stem cell differentiation and cell proliferation is a vital aim to preserve their characteristics in artificial culture and for subsequent transplantation. A number of research groups around the world developed polysaccharide templates to orchestrate stem cell activities within microcapsules. Polysaccharide hydrogels immobilize cells in three-dimensional configurations loosely controlled by the initial cell seeding density, substrate viscosity, and additives used. Their versatility and broad applicability for most tissues inside the body are significantly better than typical synthetic materials. They are also highly accessible to further development and elaboration, which increase the number of potential functions. Some of the most significant tissue responses have arisen when high-molecular-weight native extracellular matrix fragments are embedded within polysaccharide frameworks. For example, addition of human aggrecan glycosaminoglycan to biomineral-coated chitosan/alginate microcapsules accelerated the endogenous production of cartilage matrix from embedded chondroprogenitors. It is important, however, that chitosan and alginate are properly modified to optimize their biological qualities because this has vital implications for clinical outcomes. For example, it is now necessary to partially oxidize sodium alginate with sodium periodate to ensure a smooth and consistent degradation, which does not otherwise occur. The body does not naturally have enzymes that can readily breakdown alginate in its polymeric form. Another significant aspect about the precise chemistry of alginate polymer chains is the composition and arrangement of the component sugar units, guluronic acid, and mannuronic acid.

Another significant aspect about the precise chemistry of alginate polymer chains is the composition and block pattern of component sugar units. These properties have important effects on their relative bioeffectiveness. Due to their hydrophilic nature, protein deposition is more selective than on hydrophobic surfaces (more likely to recruit adhesion proteins) and can detrimentally reduce cell attachment/anchorage, which has vital implications for cell survival and the regulation of migration, proliferation, and differentiation (Rowley et al., 1999). It has been therefore important to increase cell attachment by binding peptide sequences to the alginate biopolymer or by blending candidate proteins with alginate. The sequential pattern of sugar residues along the alginate polymer chain directs and characterizes bioresponsiveness. Higher mannuronic/glucuronic (M/G) ratios lead to small pore sizes. Mannuronic-rich alginates stimulate macrophages and lymphocytes. High G alginates are preferred as they are more immunosuppressive in nature.

Another area of interest is the sustained delivery of protein and gene therapeutics. Providing the correct dosage of regenerative factors in appropriate temporal sequence has proven somewhat elusive. Gene correction strategies have evolved to overcome this significant issue. Stem cell-mediated gene therapy using nonviral transduction agents is reliant on synthetic biomaterials, lipids and physical disruption of the cell membrane to allow entry of foreign genes. However, significant cell toxicity is associated with these approaches. Natural biopolymers with transfection capacity might offer a safer alternative. Alginate–chitosan matrices have been found to yield minimal cell toxicity (less than 5%) (Green, 2004).

It is imperative that added bioactive factors are released in well-defined sequences at cell-instructive doses and for specific time periods in synchronization with the body's own biochemistry. Such actions will ensure that the added biological factors provide their maximum effect and potency. One of the main problems is releasing each factor in a slow and sustained manner for long periods of time to permanently restore tissue function. There are two ways that the release of encapsulates can be regulated. The first is to modify the composition and thickness of the capsule shell and so slow diffusion. This has been shown effective at delaying the release of plasmid deoxyribose nucleic acid. Alternatively, it is possible to create nested arrangements of beads – one inside another. The host–guest arrangement of capsules can be an effective mechanism for temporal control of encapsulates. As proof of this, exogenous release of Tyrosinase from embedded guest capsules was significantly delayed compared to Tyrosinase release from the surrounding host capsule. A further mechanism for the regulation of encapsulates is to entrap them between shell layers or within one of a succession of layers. For example, BMP-2 was entrapped at the interface between chitosan and alginate. It is possible to coat successive alternating layers of positively and negatively charged polysaccharides around the original geminal core. The versatility and adaptability of this simple-to-construct delivery system is potentially wide ranging. To exemplify, a series of bioactive proteins can be individually added inside each layer. Over time, the layers peel away from the outer surface toward the core, and release the protein encapsulates in a consecutive order. It is also then possible to engineer an elaborate series of protected domains within each shell by laminating each applied shell layer with calcium phosphate for example. Alternatively, cells can be entrapped inside the individual layers. With such assemblies, there is the capability of creating physiologically significant concentration gradients of proteins, genes, and growth factors within each polysaccharide capsule.

1.10 Concluding remarks

During the last decade, marine structures have been widely explored from the emulation of efficient designs of nature (biomimetics) to hard and soft tissue engineering, and from drug delivery to biosensor and bioreactor applications. The new directions include the use of natural organic and inorganic skeletons, nano and micro scale slow drug delivery devices, new medical treatment methods initiating unique designs and devices incorporating proteins, peptides, and stem cells.

Further studies of the way natural materials are constructed and the way they adapt to their environment will enable us to produce an exciting array of self-responsive structures and materials for regenerative medicine, structural applications, and applied engineering materials.

In nature, biomaterials are made with immaculate resource and energy efficiency using common, readily available substrates through self-assembly into highly organized hierarchies. Functional structures optimized to their environment are produced in this way.

Learning the synthesis and design methods from these marine structures will give us the opportunity to produce structures with intricate shapes and architectures that are tailored to their functions and do not break down.

We have shown previously how biomimetic approaches can yield promising outcomes for application in tissue engineering of skeletal tissues. Ongoing research toward the design of clinically relevant scaffolds for regenerative medicine using a unique set of self-organizing hierarchical structures designed and synthesized according to biological principles of design is very promising.

There is a clear and present need for better tissue engineering scaffolds that possess more natural bioresponsive environments conducive to guiding the natural processes of regeneration, which can be highly intricate and dynamic in space and time. Thus, scaffolds must have intelligence designed into them to meet this biological challenge. We contend that there needs to be a step change to scaffold environments that are responsive, whereby the synthesized biomatrix evolves in real time to meet the demands and optimizations of adaptive growth and regeneration of human tissues. As cells proliferate and differentiate, they alter their environment.

Future advanced biomimetic scaffolds must be able to adapt to these changes and meet the ever-changing needs of developing tissues. We anticipate synthesizing biomaterial scaffolds with functional cross-links and pendant side groups that interact with surrounding cell population at three levels: at the contact interface (nano/meso level), at the architectural (micro level), and at the biofunctional (macro level).

Nanofabrication using biological principles of assembly and design is still in its infancy. Use of this bioinspired nanofabrication for tissue engineering is a unique approach that has enormous potential to improve scaffold design and tailor physico-chemical environments with an ability to microevolve. Such revolutionary technology will also engender benefit in materials science and its application. Synthetic materials with characteristics and function of biological ones are sought by structural, aerospace, and automotive engineers, for example. So far, the techniques described and that are being developed now are made without direct cell guidance, under no genetic control. This is the next challenge: to grow materials with cells and to promote their regulation of material synthesis.

Materials with capacity for self-actuation, response, and repair are also sought by construction and defense industries. Biomaterials constantly adapt their composition, growth and hence function to dynamically changing environmental conditions and have the ability for self-repair when adaptation fails or is too slow. Advanced functional materials with these characteristics would see enormous benefit to engineering industries and biomedicine. Marine structures with their unique designs, chemistry, adaptability, and functionality with most efficient ways offer an excellent learning platform. At a time when using fewer sustainable resources, consuming less energy, and efficiency in design and production routes, this knowledge is such an imperative.

Acknowledgments

The author would like to thank to a large number of Ph.D. students and postdoctoral fellows contributed to advancement of marine structure-based biomaterials and the related experimental work. Special thanks to Prof. B. Milthorpe, A/Prof. D.W. Green, and Dr J. Chou for their continued support in search of the most appropriate biomaterial.

References

Abramovitch-Gottlib, L., Geresh, S., Vago, R., 2006. Biofabricated marine hydrozoan: a bioactive crystalline material promoting ossification of mesenchymal stem cells. Tissue Eng. 12 (4), 729–739.

Aizenberg, J., Hendler, G., 2004. Designing efficient microlens arrays: lessons from nature. J. Mater. Chem. 14, 2066–2072.

Almeida, M.J., Pereira, L., Milet, C., Haigle, J., Barbosa, M., Lopez, E., 2001. Comparative effects of nacre water-soluble matrix and dexamethasone on the alkaline phosphatase activity of MRC-5 fibroblasts. J. Biomed. Mater. Res. 57, 306–312.

Ando, J., 1958. Tricalcium phosphate and its variation. Bull. Chem. Soc. Jpn. 31, 196–201.

Andrianantoandro, E., Basu, S., Karig, D.K., Weiss, R., 2006. Synthetic biology: new engineering rules for an emerging discipline. Mol. Syst. Biol. 2, 2006.0028, Epub 2006 May 16.

Bahar, H., Yaffe, A., Binderman, I., 2003. The influence of nacre surface and its modification on bone apposition: a bone development model in rats. J. Periodontol. 74, 366–371.

Belegratis, M.R., Schmidt, V., Nees, D., Stadlober, B., Hartmann, P., 2014. Diatom-inspired templates for 3D replication: natural diatoms versus laser written artificial diatoms. Bioinspir. Biomim. 9 (1), 016004. http://dx.doi.org/10.1088/1748-3182/9/1/016004.

Ben-Nissan, B., 2003. Natural bioceramics: from coral to bone and beyond. Curr. Opin. Solid State Mater. Sci. 7 (4–5), 283–288.

Ben-Nissan, B., 2004. Biomimetics and bioceramics. In: Reis, R.L., Weiner, S. (Eds.), Learning from Nature How to Design New Implantable Biomaterials. NATO Science Series, vol. 171. Academic Publishers, New York, pp. 89–103.

Ben-Nissan, B., Green, D.W., 2013. Marine materials in drug delivery and tissue engineering: from natural role models, to bone regeneration and repair and slow delivery of therapeutic drugs, proteins and genes. In: Marine Biomaterials. Taylor and Francis, CSR Books, Boca Raton, FL, pp. 575–602, Chapter 30.

Bonnelye, E., Chabadel, A., Saltel, F., Jurdic, P., 2008. Dual effect of strontium ranelate: stimulation of osteoblast differentiation and inhibition of osteoclast formation and resorption in vitro. Bone 42 (1), 129–138.

Boute, N., Exposito, J.-Y., Boury-Esnault, N., Vacelet, J., Noro, N., Miyazaki, K., Yoshizato, K., Garrone, R., 1996. Type IV collagen in sponges, the missing link in basement membrane ubiquity. Biol. Cell 88, 37–44.

Choi, A.H., Ben-Nissan, B., 2007. Sol–gel production of bioactive nanocoatings for medical applications. Part II: current research and development. Nanomedicine 2 (1), 51–61.

Chou, J., Ben-Nissan, B., Green, D.W., Valenzuela, S.M., Kohan, L., 2011. Targeting and dissolution characteristics of bone forming and antibacterial drugs by harnessing the structure of micro-spherical shells from coral beach sand. Adv. Eng. Mater. 13 (1–2), 93–99.

Chou, J., Valenzuela, S., Bishop, D., Ben-Nissan, B., Milthorpe, B., 2012. Strontium- and magnesium-enriched biomimetic beta-TCP macrospheres with potential for bone tissue morphogenesis. J. Tissue Eng. Regen. Med. http://dx.doi.org/10.1002/term.1576.

Chou, J., Ito, T., Bishop, D., Otsuka, M., Ben-Nissan, B., Milthorpe, B., 2013. Controlled release of simvastatin from biomimetic β-TCP drug delivery system. PLoS ONE 8 (1), 1–6.

Dawkins, R., 1989. The evolution of evolvability. In: Langton, C. (Ed.), Artificial Life. Addison-Wesley, Reading, MA.

Duplat, D., Chabadel, A., Gallet, M., Berland, S., Bedouet, L., Rousseau, M., et al., 2007. The in vitro osteoclastic degradation of nacre. Biomaterials 28 (12), 2155–2162.

Ehrlich, H., Etnoyer, P., Litvinov, S.D., Olennikova, M.M., Domaschke, H., Hanke, T., et al., 2006. Biomaterial structure in deep-sea bamboo coral (Anthozoa: Gorgonacea: Isididae): perspectives for the development of bone implants and templates for tissue engineering. Materialwiss. Werkstofftech. 37 (6), 552–557.

Elliott, J.C., 1994. Structure and Chemistry of the Apatites and Other Calcium Orthophosphates. Elsevier, New York, p. 44.

Exposito, J.Y., Cluzel, C., Garrone, R., Lethias, C.M., 2002. Evolution of collagens. Anat. Rec. 268, 302–316.

Gong, Y.U.T., Killian, C.E., Olson, I.C., Appathurai, N.P., Amasino, A.L., Martin, M.C., Holt, L.J., Wilt, F.H., Gilbert, P.U.P.A., 2012. Phase transitions in biogenic amorphous calcium carbonate. Proc. Natl Acad. Sci. USA. 109 (16), 6088–6093. http://dx.doi.org/10.1073/pnas.1118085109.

Gonzalez-McQuire, R., Green, D., Walsh, D., Hall, S.R., Chane-Ching, J.-Y., Oreffo, R.O.C., Mann, S., 2005. Fabrication of hydroxyapatite sponges by dextran sulfate/amino acid templating. Biomaterials 26, 6652–6656.

Green, D.W., 2004. Bio-inspired ceramic structures: from invertebrate marine skeletons to biomimetic crystal engineering. J. Aust. Ceram. Soc. 40 (2), 1–6.

Green, D.W., Ben-Nissan, B., 2008. Bio-inspired engineering of human tissue scaffolding in regenerative medicine. In: Tateishi, T. (Ed.), Biomaterials in Asia. World Scientific, Singapore, pp. 364–388. ISBN: 978-981-283-574-1.

Green, D.W., Ben-Nissan, B., 2010. Biomimetic applications in regenerative medicine: scaffolds, transplantation modules, tissue homing device and stem cells. Chapter 21. In: Tochilin, V., Amiji, M. (Eds.), Handbook of Materials for Nanomedicine. Pan Stanford Publishing Pte Ltd, Singapore, pp. 821–850. ISBN: 978-981-4267-55-7.

Green, D., Walsh, D., Yang, X., Mann, S., Oreffo, R.O.C., 2004. Stimulation of human bone marrow stromal cells using growth factor-encapsulated calcium carbonate porous microspheres. J. Mater. Chem. 14, 2206–2212.

Green, D.W., Li, G., Milthorpe, B., Ben-Nissan, B., 2012. Adult stem cell coatings using biomaterials for regenerative medicine. Mater. Today 15 (1–2), 61–68.

Green, D.W., Padula, M.P., Santos, J., Chou, J., Milthorpe, B., Ben-Nissan, B., 2013. A therapeutic potential for marine skeletal proteins in bone regeneration. Mar. Drugs 11 (4), 1203–1220.

Grunert, M., Nurcombe, V., Cool, S.M., 2008. Stem cell fate decisions: the role of heparan sulfate in the control of autocrine and paracrine signals. Curr. Stem Cell Res. Ther. 3 (1), 1–8.

Hall, S.R., Swinerd, V.M., Newby, F.N., Collins, A.M., Mann, S., 2006. Fabrication of porous titania (brookite) microparticles with complex morphology by sol–gel replication of pollen grains. Chem. Mater. 18, 598–600.

Hu, J., Russell, J.J., Ben-Nissan, B., Vago, R., 2001. Production and analysis of hydroxyapatite from Australian corals via hydrothermal process. J. Mater. Sci. Lett. 20 (1), 85–87.

Ingber, D.E., 1998. The architecture of life. Sci. Am. 278, 48–57.

Kel, D., Gökçe, H., Bilgiç, D., Ağaoğulları, D., Duman, I., Öveçoğlu, M.L., et al., 2012. Production of natural HA from land snails. Key Eng. Mater. 493, 287–292.

Vecchio, K.S., Zhang, X., Massie, J.B., Wang, M., Kim, C.W., 2007. Conversion of sea urchin spines to Mg-substituted tricalcium phosphate for bone implants. Acta Biomater. 3, 785–793.

Kim, E.S., 2008. Directed evolution: a historical exploration into an evolutionary experimental system of nanobiotechnology, 1965–2006. Minerva 46, 463–484.

Kim, Y.M., Kim, J.J., Kim, Y.H., Rho, J.Y., 2000. Effects of organic matrix proteins on the interfacial structures at the bone-biocompatible nacre interface in vitro. Biomaterials 23, 2089–2096.

Lamghari, M., Berland, S., Laurent, A., Huet, H., Lopez, E., 2001. Bone reactions to nacre injected percutaneously into the vertebrae of sheep. Biomaterials 22, 555–562.

LeGeros, R., 1981. Apatites in biological systems. Progr. Cryst. Growth Charact. Mater. 41 (1–2), 1–45.

LeGeros, R.Z., 1993. Biodegradation and bioresorption of calcium phosphate ceramics. Clin. Mater. 4, 65–88.

LeGeros, R.Z., Gatti, A.M., Kijkowska, R., Mijares, D.Q., LeGeros, J.P., 2004. Magnesium tricalcium phosphate: formation and properties. Key Eng. Mater. 254–256, 127–130.

Leupold, J., Barfield, W., An, Y., Hartsock, L., 2006. A comparison of ProOsteon, DBX, and collagraft in a rabbit model. J. Biomed. Mater. Res. B Appl. Biomater. 79 (2), 292–297.

Liao, H., Mutvei, H., Hammarstrom, L., Wurtz, T., Li, J., 2002. Tissue responses to nacreous implants in rat femur: an in situ hybridization and histochemical study. Biomaterials 23, 2693–2701.

Lopez, E., Vidal, B., Berland, S., Camprasse, S., Camprasse, G., Silve, C., 1992. Demonstration of the capacity of nacre to induce bone formation by human osteoblasts maintained *in vitro*. Tissue Cell 24, 667–679.

Macha, I.J., Ozyegin, L.S., Chou, J., Samur, R., Oktar, F.N., Ben-Nissan, B., 2013. An alternative synthesis method for dicalcium phosphate (monetite) powders from Mediterranean mussel (*Mytilus galloprovincialis*) shells. J. Aust. Ceram. Soc. 49 (2), 122–128.

Mann, S., 1983. Mineralization in biological systems. Struct. Bond. 54, 125.

Mann, S., 2001. Biomineralization principles and concepts in bioinorganic materials chemistry. Oxford University Press, New York, NY.

Marga, F., Neagu, A., Kosztin, I., Forgacs, G., 2007. Developmental biology and tissue engineering. Birth Defects Res. C 81 (4), 320–328.

Martina, M., Subramanyam, G., Weaver, J.C., Hutmacher, D.W., Morse, D.E., Valiyaveettil, S., 2005. Developing macroporous bicontinuous materials as scaffolds for tissue engineering. Biomaterials 26, 5609–5616.

Milev, A.S., Kannangara, G.S.K., Ben-Nissan, B., 2003. Morphological stability of plate-like carbonated hydroxyapatite. Mater. Lett. 57 (13–14), 1960–1965.

Nicklas, M., Schatton, W., Heinemann, S., Hanke, T., Kreuter, J., 2009. Preparation and characterization of marine sponge collagen nanoparticles and employment for the transdermal delivery of 17β-estradiol-hemihydrate. Drug Develop. Ind. Pharma 35 (9), 1035–1042.

Nowak, D., Florek, M., Nowak, J., Kwiatek, W., Lekki, J., Zieba, E., et al., 2005. Microspectrometric investigations of inorganic components of the black corals for biomedical applications. Key Eng. Mater. 284–286, 297–300.

Nurcombe, V., Cool, S.M., 2007. Heparan sulfate control of proliferation and differentiation in the stem cell niche. Crit. Rev. Eukaryot. Gene Expr. 17 (2), 159–171.

Papacharalambous, S., Anastasoff, K., 1993. Natural coral skeleton used as onlay graft for contour augmentation of the face. A preliminary report. Int. J. Oral Maxillofac. Surg. 22 (5), 260–264.

Parker, A.R., Townley, H.E., 2007. Biomimetics of photonic nanostructures. Nat. Nanotechnol. 2, 347–353.

Petite, H., Kacem, K., Triffitt, J.T., 1996. Adhesion, growth and differentiation of human bone marrow cells on non porous calcium carbonate and pastic substrata: effects of dexamethasone and 1,25 dihydroxyvitamin D3. Mater. Med. 7, 665–671.

Politi, Y., Arad, T., Klein, E., Weiner, S., Addadi, L., 2004. Sea urchin spine calcite forms via a transient amorphous calcium carbonate phase. Science 306, 1161–1164.

Raz, S., Hamilton, P.C., Wilt, F.H., Weiner, S., Addadi, L., 2003. The transient phase of amorphous calcium carbonate in sea urchin larval spicules: the involvement of proteins and magnesium ions in its formation and stabilization. Adv. Funct. Mater. 13, 480–486.

Rocha, J.H.G., Lemos, A.F., Agathopoulos, V.P., Kannan, S., Oktar, F.N., Ferriera, J.M.F., 2005. Scaffolds for bone restoration from cuttlefish. Bone, 37, 850–857.

Rocha, J.H.G., Lemos, A.F., Agathopoulos, S., Kannan, S., Valerio, P., Ferreira, J.M.F., 2006. Hydrothermal growth of hydroxyapatite scaffolds from aragonite cuttlefish bones. J. Biomed. Mater. Res. A 77, 160–168.

Rousseau, M., Lucilia, P.M., Almeida, M.J., Milet, C., Lopez, E., 2003. The water-soluble matrix fraction from the nacre of *Pinctada maxima* produces earlier mineralization of MC3T3-E1 mouse pre-osteoblasts. Comp. Biochem. Physiol. B 135, 1–7.

Rousseau, M., Boulzaguet, H., Biagianti, J., Duplat, D., Milet, C., Lopez, E., et al., 2007. Low molecular weight molecules of oyster nacre induce mineralization of the MC3T3-E1. J. Biomed. Mater. Res. A 85A, 487–497.

Roy, D.M., Linnehan, S., 1974. Hydroxyapatite formed from coral skeletal carbonate by hydrothermal exchange. Nature 247, 220–222.

Rowley, J.A., Madlambayana, G., Mooney, D.J., 1999. Alginate hydrogels as synthetic extracellular matrix materials. Biomaterials 20 (1), 45–53.

Samur, R., Ozyegin, L., Agaogullari, N., Oktar, F.N., Agatopoulos, S., Kalkandelen, C., et al., 2013. Calcium phosphate formation from sea urchin (*Brissus latecarinatus*) via modified mechano chemical (ultrasonic) conversion method. Metallurgia 52 (3), 375–378.

Sanchez, C., Arribart, H., Madeleine, M., Guille, G., 2007. Biomimetism and bioinspiration as tools for the design of innovative materials and systems. Nat. Mater. 4, 277–287.

Sengers, B.G., Please, C.P., Oreffo, R.O.C., 2007. Experimental characterization and computational modeling of two-dimensional cell spreading for skeletal regeneration. J. R. Soc. Interface 4, 1107–1117.

Shen, Y., Zhu, J., Zhang, H., Zhao, F., 2006. In vitro osteogenic activity of pearl. Biomaterials 27, 281–287.

Stanley, G., 2003. The evolution of modern corals and their early history. Earth Sci. Rev. 60 (3–4), 195–225.

Su, X., Kamat, S., Heuer, A.H., 2000. The structure of sea urchin spines, large biogenic single crystals of calcite. J. Mater. Sci. 35, 5545–5551.

Swatschek, D., Schatton, W., Kellermann, J., Muller, W.E.G., Kreuter, J., 2002. Marine sponge collagen: isolation, characterization and effects on the skin parameters surface pH, moisture and sebum. Eur. J. Pharma Biopharm. 53, 107–113.

Townley, H.E., Parker, A.R., White-Cooper, H., 2008. Exploitation of diatom frustules for nanotechnology: tethering active biomolecules. Adv. Funct. Mater. 18, 369–374.

Vago, R., 2008. Beyond the skeleton. Cnidarian biomaterials as bioactive extracellular microenvironments for tissue engineering. Organogenesis 4 (1), 18–22.

Vago, R., Plotquin, D., Bunin, A., Sinelnikov, I., Atar, D., Itzhak, D., 2002. Hard tissue remodeling using biofabricated coralline biomaterials. J. Biochem. Biophys. Methods 50, 253–259.

Walsh, D., Mann, S., 1996. Feigning nature's sculptures. Chem. Brit. 32, 31–34.

Walsh, D., Boanini, E., Tanaka, J., Mann, S., 2005. Synthesis of tri-calcium phosphate sponges by interfacial deposition and thermal transformation of self-supporting inorganic films. J. Mater. Chem. 15, 1043–1048.

Weber, J.N., Greer, R., Voight, B., White, E., Roy, R., 1969. Unusual strength properties of echinoderm calcite related. J. Ultrastruct. Res. 26, 355–366.

Weber, J.N., White, E.W., Lebiedzik, J., 1971. New porous biomaterials by replication of echinoderm skeletal microstructures. Nature 233, 2337–2339.

Westbroek, P., Marin, F., 1998. A marriage of bone and nacre. Nature 392, 861–862.

White, R.A., Weber, J.N., White, E.W., 1972. Replamineform: a new process. Science 176, 92.

Williams, G.C., Nesse, R.M., 1991. The dawn of Darwinian medicine. Quart. Rev. Biol. 66 (1), 1–22.

Zhang, C., Li, S., Ma, Z., Xie, L., Zhang, R., 2006. A novel matrix protein p10 from the nacre of pearl oyster (*Pinctada fucata*) and its effects on both $CaCO_3$ crystal formation and mineralogic cells. Mar. Biotechnol. 8, 624–633.

Properties and applications of biosilica enzymatically synthesized by aquatic/marine sponges

2

X. Wang, H.C. Schröder, W.E.G. Müller

2.1 Introduction

Sponges (phylum Porifera) are aquatic, sessile, and multi-cellular organisms with a Bauplan that appears simple at a first glance and lacks similarities to any other living organism. Therefore, during early studies, it was difficult to determine morphological characters that would conclusively allow to group sponges into either one of the two kingdoms of multi-cellular life: Metazoa or Plantae. After comprehensive isolation, cloning, and phylogenetic analyses of many poriferan genes by our group, it became obvious that the phylum Porifera comprises three classes – Hexactinellida, Demospongiae, and Calcarea – and forms the basis of the metazoan kingdom (Müller, 1995). A few years later, it could be clarified that Hexactinellida (glass sponges), Demospongiae (silicate/spongin sponges), and Calcarea (calcareous sponges) are monophyletic and closely related to the common ancestor of all metazoans, the Urmetazoa (Müller, 2001).

Sponges appeared during the Neoproterozoic era, the geologic period from 1000 to 542 Ma (reviewed in Müller et al., 2007a). Two major reasons contributed to the evolutionary success of the poriferan taxon: (a) symbiosis with microorganisms and (b) presence of hard skeletons (Müller et al., 2007a). The maintenance of symbiotic relationships with unicellular organisms allowed sponges to survive adverse environmental conditions because the autotrophic microbial symbionts represented rich organic carbon sources. On the other hand, the development of skeletal elements facilitated an increase in size, a common metazoan phyletic trend also known as Cope's rule (Nicol, 1966): Since changes in body size affect almost every aspect of life (Schmidt-Nielsen, 1984), two strategies have been developed in animals to circumvent any constraints (reviewed in Page, 2007), first by acquisition of a hydrostatic skeleton, as it is known from the "worm"-like phyla of the Ediacaran and pre-Ediacaran Eon (Xiao and Kaufman, 2006), or second by acquisition of rigid solid skeletal elements (Alexander et al., 1979; Biewener, 2005), as they were realized in Neoproterozoic siliceous sponges (see Müller et al., 2007a).

Skeletal elements (spicules) of siliceous sponges, Hexactinellida and Demospongiae, are composed of amorphous opal ($SiO_2 \cdot nH_2O$). They already existed in pre-Ediacaran sponges and represent a general and basic morphological character until today (Xiao et al., 2000). It is easily conceivable why the animals integrated silicon instead of calcium as the fundamental element for their inorganic skeleton, since the

Functional Marine Biomaterials. http://dx.doi.org/10.1016/B978-1-78242-086-6.00002-9
Copyright © 2015 Elsevier Ltd. All rights reserved.

Neoproterozoic oceans were rich in silicic acid and continuously replenished by products of the silicate weathering-carbonate precipitation cycle (Walker, 2003). Sponges are provided with a sophisticated aquiferous canal system and hence, their body can be stabilized to some extent by internal hydrostatic pressure, even though a vasculomuscular system does not exist. The presence of a siliceous skeleton, however, allowed them to obtain sizes larger than 20 mm, even up to 2.5 m (Wang et al., 2009).

2.2 Silicatein-based siliceous spicule formation

The biogenic basis of spicule formation and the turnover of silica in spicules have already been depicted by Duncan (1881). He formulated "The spicule which has lived, has to decay, and may live again in another form." However, it took until 1999 until Cha et al. discovered that the main constituent of the proteinaceous filament within the axial canal of spicules is an enzyme, subsequently termed silicatein, which might be involved in biosilica formation. Soon after having identified this anabolic enzyme, also the corresponding catabolic enzyme (silicase) was discovered (Schröder et al., 2003). The identification of a biosilica degrading enzyme supported the view that the siliceous components in spicules are under metabolic turnover (Eckert et al., 2006). Studies on the metabolism of spicules on the cellular level became possible after the introduction of a poriferan cell culture system, primmorphs (Imsiecke and Müller, 1995; Custódio et al., 1998). Already the first contribution on that topic resolved that spicule formation starts intracellularly in "J" sclerocytes, by formation of an initial organic axial filament, around which the inorganic silica mantel is deposited. This result had later been corroborated by application of more advanced immunochemical and electron microscopy techniques (Müller et al., 2007b).

2.2.1 Silicatein

Thorough dissolution of the inorganic silica from spicules (Figure 2.1d; *Suberites domuncula* [demosponge] spicules) with hydrofluoric acid (HF) vapor revealed in addition to the presence of the axial filament a spicule-enfolding proteinaceous coat (Figure 2.1e). The siliceous shell of the spicules can be dissolved by exposure to HF. During this process, the polymerized silicic acid undergoes dissolution into monomeric silicic acid under the release of the organic component(s) of the spicules. Focusing on the tylostyles from the demosponge *S. domuncula*, both the knob of the spicules (Figure 2.1d and e) and the axial rod (Figure 2.1f) undergo dissolution under release of the organic axial filament, located within the axial canal, and additional proteins that are obviously initially localized within the silica coat of the spicule (Figure 2.1d–f). While in the immature spicules the axial filament fills the complete axial canal, it shrinks at later stages (Figure 2.1a–c). A pile of isolated axial filaments is shown in Figure 2.1I.

After the discovery of the cathepsin L (cysteine protease)-related silicatein (Shimizu et al., 1998; Cha et al., 1999) in spicules of the demosponge *Tethya aurantium*, several

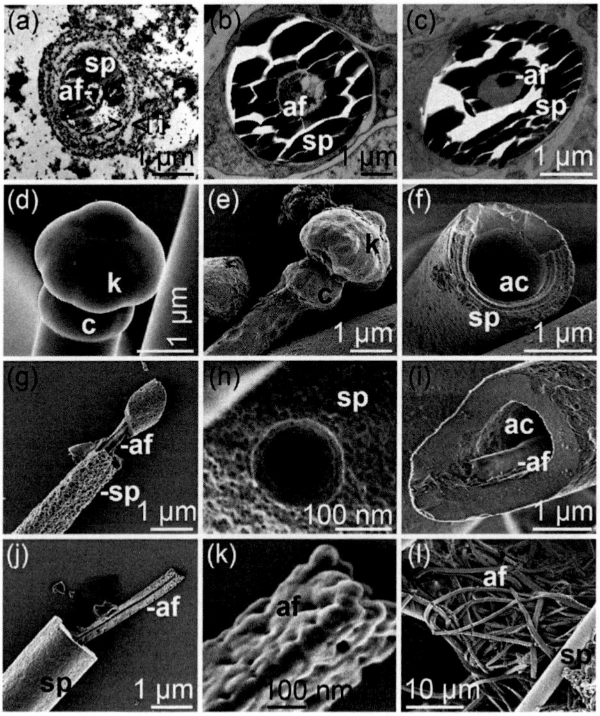

Figure 2.1 Localization of the silicatein, packed within the axial filament, in the spicules from *S. domuncula*. (a) After rising of antibodies to silicatein, it became possible to demonstrate, by electron immunogold labeling technique, that concentric rings (ri) surround the growing spicules (sp) and react also with the axial filament (af). This approach also showed that the silicatein molecules are attached to the string- and net-like structures. (b) Immature spicule harboring a space-filling axial filament (af) within the axial canal. (c) In a more mature spicule (sp), the axial filament (af) had been shrunk. (d) A tylostyle head, the knob (k), is based on a collar (c) at one end of the monaxonal pin. HR-SEM: (e) After exposure of the tylostyle to HF vapor, the silica is progressively dissolved, exposing granular structures. (f) A broken *S. domuncula* spicule (sp), a tylostyle, displaying the axial canal (ac). (g to j) Progressive dissolution of the siliceous shell of the spicules (sp) releasing the axial filament (af). (h) At a higher magnification, the plait cable-like organization of the axial filament (af) becomes obvious. (k and l) Bundles of axial filaments (af) released after HF treatment.

related genes were elucidated in both marine and freshwater demosponges (reviewed in Müller et al., 2007b). The corresponding deduced polypeptides comprise about 325 amino acids (aa) with a molecular weight of ca. 35 kDa. During maturation, this primary translation product (proenzyme) is processed by cleaving off a signal peptide (aa_1–aa_{17}; *S. domuncula* [demosponge] silicatein-α) and the adjacent propeptide (aa_{18}–aa_{112}), resulting in the mature enzyme that has a size of 24–25 kDa. Similar to cathepsins, the catalytic center of silicatein contains His and Asn. However, the Cys of the cathepsins' catalytic triad is exchanged by Ser in silicatein. In addition to about 10 putative protein kinase phosphorylation sites, silicateins display a cluster of serine residues that is found close to the central aa residue of the catalytic triad, but is otherwise missing in cathepsins.

Subsequent phylogenetic analyses revealed that silicateins form a separate branch from cathepsins (Müller et al., 2007b; Figure 2.2). The difficult accessibility of hexactinellids, which live primarily in depths of more than 300 m, generally results in a very poor sampling. Accordingly, only recently the first hexactinellid silicatein (*Crateromorpha meyeri*) could be identified and characterized (Müller et al., 2008a). This molecule shares high similarity to the demosponge sequences (expect value of 8×10^{-58}) and contains the same catalytic triad amino acids. However, striking in the *C. meyeri* sequence is a second Ser-rich cluster, which is located between the second and the third aa of the catalytic triad. Strong binding of the protein to the spicule silica surface has been attributed to this cluster (Müller et al., 2008b). The posttranslational modifications of silicatein have been found to be essential for the enzyme activity with respect to (a) association with other structural and functional molecules within the tissue and (b) self-association/self-assembly. For those studies, silicatein had been isolated from spicules in the absence of HF, but in the presence of a glycerol-based buffer. Following this rationale, it could be demonstrated that silicatein exists not only in the axial canal but also in the extraspicular and extracellular space (Müller et al., 2005; Schröder et al., 2006). The enzymatic reaction mechanism of silicatein had been proposed by Cha et al. (1999); the detailed properties of the reaction kinetics have been specified experimentally (Müller et al., 2008c).

2.3 Spiculogenesis

With the primmorph cell culture system, the first solid evidence was provided that the synthesis of the spicules starts intracellularly (Müller et al., 2005). The process of spicule formation can be divided into an initial intracellular phase and a subsequent extracellular shaping phase. Silicic acid, the substrate for silicatein, is actively taken up by cells (sclerocytes) via the Na^+/HCO_3^- ($Si(OH)_4$) cotransporter. In parallel, silicatein is synthesized, processed, and stored with silicic acid in special organelles of the sclerocytes, the silicasomes. Within these organelles, the axial filaments are formed, around which silica is deposited enzymatically.

After formation of a first biosilica layer, immature spicules are released into the extracellular space, where they grow in length (axial/longitudinal direction) and in diameter (radial direction), by appositional layering of silica lamellae (Schröder et al., 2006). During growth in the axial direction, biosilica is formed through the enzymatic function of the axial filament silicateins. The growth of the spicule is driven by the elongating biosilica core cylinder, which is synthesized by the 23-kDa processed form of silicatein. The radial thickening of the spicules, their appositional growth (radial direction), occurs by deposition of silica on the surface of the growing spicule, and is mediated by the extraspicular 34.7-kDa immature silicatein. There is no evidence at all that, in demosponges, either in the axial filament or on the surface of the spicules, there is collagen that is causatively involved in biosilica formation.

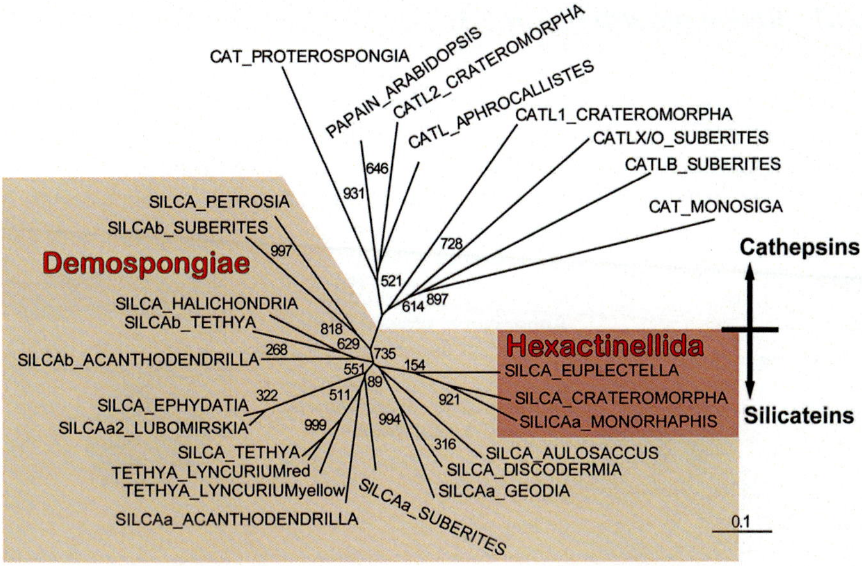

Figure 2.2 Phylogenetic analysis of silicatein within the cathepsin family. The deduced proteins were aligned and the phylogenetic tree was constructed. The hitherto-known four hexactinellid sequences were included: silicatein from *Euplectella aspergillum* (SILCA_EUPLECTELLA; FR748156), *C. meyeri* (SILCA_CRATEROMORPHA; AM920776), *Monorhaphis chuni* (SILCAa_MONORHAPHIS; FN394978), and the silicatein-like protein *Aulosaccus* sp. (SILCA_AULOSACCUS; ACU86976.1). The bulk of silicatein sequences have been identified in demosponges. First, the silicatein-α sequences from *S. domuncula* (SILCAa_SUBERITES; CAC03737.1), *T. aurantium* (SILCAa_TETHYA; AAC23951.1), *Tethya lycurium* red variation (TETHYA_LYCURIUMred; FR748154) and *T. lycurium* yellow variation (TETHYA_LYCURIUMyellow; FR748155), *Discodermia japonica* (SILCA_DISCODERMiA; FR748157), *Geodia cydonium* (SILCAa_GEODIA; CAM57981.1) and *Acanthodendrilla* sp. Vietnam (SILCAa_ACANTHODENDRILLA; ACH92669.1), as well as from *Lubomirskia baicalensis* (SILCAa2_LUBOMIRSKIA; AJ968945) and *Ephydatia fluviatilis* (SILCA_EPHYDATIA; BAE54434.1). Second, the silicatein-β sequences from *S. domuncula* (SILCAb_SUBERITES; CAH04635.1), *T. aurantium* (SILCAb_TETHYA; AF098670_1), and *Acanthodendrilla* sp. Vietnam (SILCAb_ACANTHODENDRILLA; FJ013043.1). Third, the silicateins that had been identified in marine sponges from which only one isoform had been obtained; silicatein from *Petrosia ficiformis* (SILCA_PETROSIA; AAO23671.1) and *Halichondria okadai* (SILCA_HALICHONDRIA; BAB86343.1). As reflected in the rooted tree, these silicateins derived from the cathepsins among which in this tree the following sequences have been included: cathepsin-like protein 2 *C. meyeri* (CATL2_CRATEROMORPHA; CAP17585.1), cathepsin-like protein 1 *C. meyeri* (CATL1_CRATEROMORPHA; CAP17584.1), mRNA for cathepsin L (catl gene) *Aphrocallistes vastus* (CATL_APHROCALLISTES; AJ968951), cathepsin B *S. domuncula* (CATLB_SUBERITES; CAH04630.1), and cathepsin X/O *S. domuncula* (CATLX/O_SUBERITES; CAH04633.1). The resulting tree was rooted with the sequence from the papain-like cysteine peptidase XBCP3 *Arabidopsis thaliana* (PAPAIN_ARABIDOPSIS; AF388175_1). In addition, the cathepsins from choanoflagellates had been included to show that they were derived, according to this tree, from the sponge cathepsins; the cysteine protease from *Proterospongia* sp. (PRL00000040) and the cathepsin from *Monosiga ovata* MNL00000103.

2.3.1 *Radial growth* (*Figure 2.3*)

The radial, appositional thickening of the spicules and then the axial elongation process (Schröder et al., 2006) was first understood. In the extracellular space, the silicatein molecules are organized into larger entities, by concentric rings/cylinders around the spicule surface. These structures become stabilized by the protein galectin and Ca^{2+}; within these cylinders, the silicatein-mediated biosilica formation occurs. In

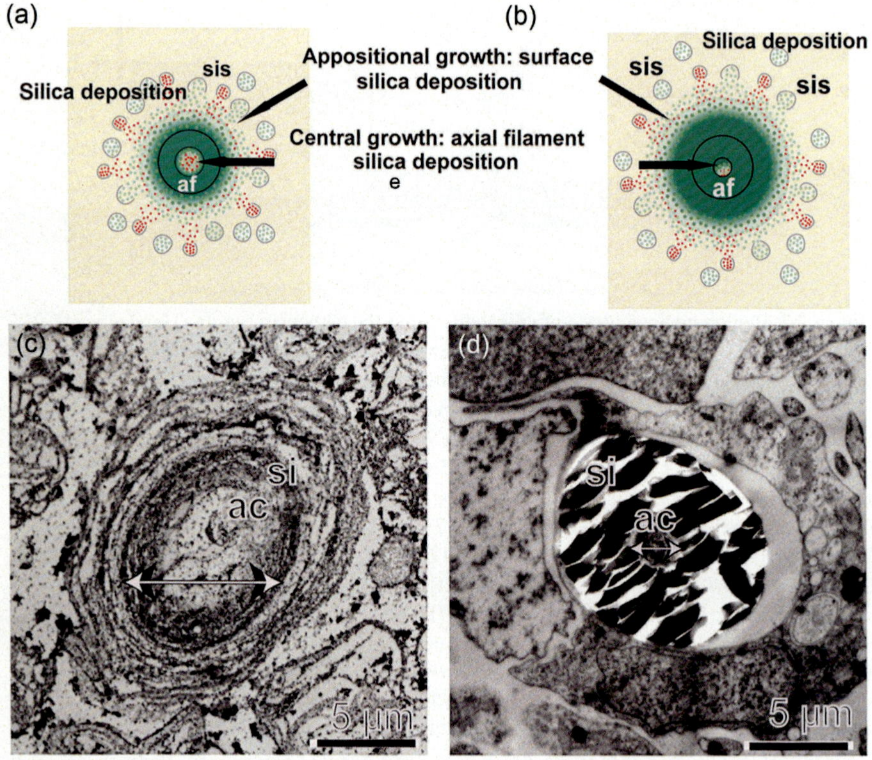

Figure 2.3 The growth of the spicules occurs in two directions. This scheme shows the process of radial thickening of the spicule. (a and b) In radial direction, the spicules increase in thickness by appositional layering of the growing spicule (scheme). Both within the axial canal (ac), where the axial filament (formed of silicatein) is located (central growth), and on the surface of the spicules (appositional growth), the silica shell (si) increases in dimension through enzymatic synthesis of biosilica via silicatein. In the extra-spicular space, both silicatein and silica are released from silicasomes (sis); those silicasomes are set free from the surrounding cells. (c and d) Shrinkage of biosilica during biosilicification. (c) Growing siliceous spicules comprise a telescoped cylindrical arrangement. Approximately seven cylinders, formed from galectin, surround the axial canal (ac). Silicification, reflected by the densely packed grana, proceeds progressively from the central cylinders toward the surface of the spicule. TEM, immunogold labeling with anti-silicatein antibodies. (d) Section through a mature spicule showing the axial canal in the center. The shrinking both of the spicule during formation/maturation and of the axial canal is obvious.

hexactinellid spicules, these appositionally layered silica lamellae remain separated, and can reach 1000 in number (Wang et al., 2009). In contrast, in demosponges, the individual lamellae fuse/bio-sinter together, and form a “solid” siliceous shell that surrounds the centrally located axial filament.

2.3.2 Longitudinal growth *(Figure 2.4)*

We provided, for the first time, experimental evidence for the involvement of cellular processes in the control of the axial growth of spicules. Using the primmorph system, we demonstrated that these cell processes originate from evaginations of the

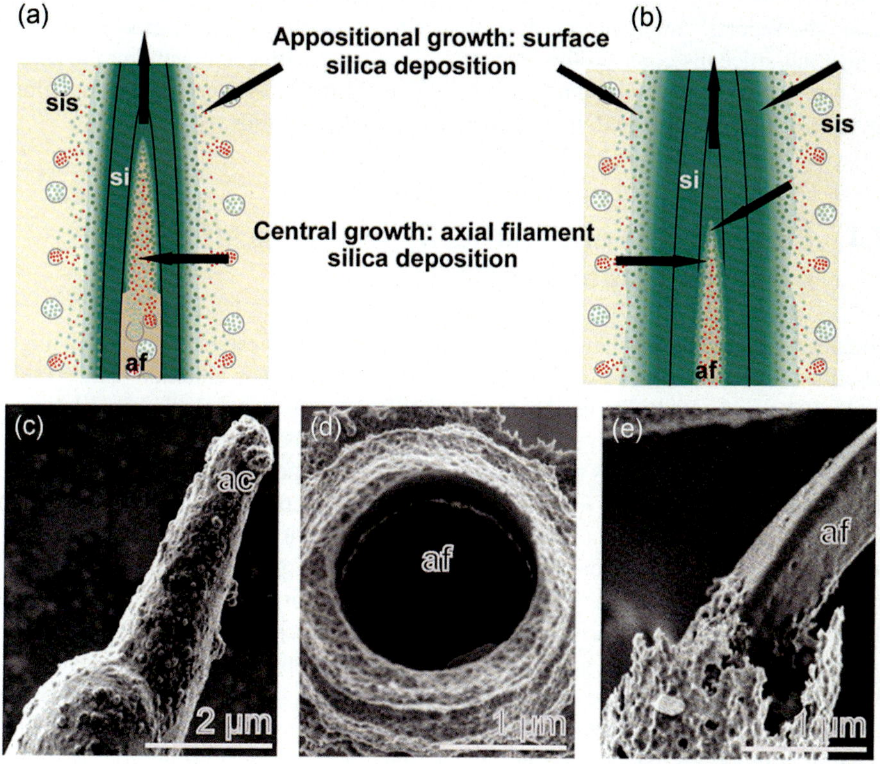

Figure 2.4 Longitudinal growth of the spicules. (a and b) After the release of the immature spicule into the extracellular space via evagination of a cell protrusion, the growing spicule is pushed away from the cell (sclerocyte) that initially has formed the immature spicule (scheme). This process is driven by an elongation of the cell protrusion and in turn an elongation of the axial filament (af) within the axial canal. Again, it is highlighted that silica deposition occurs within the axial canal and onto the surface of the spicules. Around the spicules, silicasomes (sis) exist, which release silicatein and orthosilicate to allow biosilica formation to occur. (c) A protruding axial canal (ac) is shown that harbors an axial filament. (d) Silica shell around the axial canal/axial filament (af). (e) A liberated axial filament (af) projecting from a spicule.

spicule-forming cells (sclerocytes) into the growing and elongating axial canal. The experiments showed that, around a cell extension that protrudes from a sclerocyte into the axial canal of a given spicule, silicatein molecules are released into the extracellular space of the axial canal (the space between the cell membrane and the inner surface of the siliceous mantel) to catalyze biosilica deposition from the inner surface. This causes the axial canal to narrow from >1 to <0.5 μm. Intracellularly, both the enzyme and its substrate (silica precursor) are stored in vesicles that have been termed silicasomes (Müller et al., 2009a,b). These vesicles are released into the axial canal to allow the enzymatic polycondensation reaction.

Therefore, spicule formation requires two mechanistically independent biosilica condensation/deposition processes: first, biosilica formation, which gives rise to the inner core of the growing spicule, mediated by silicatein present in the axial filament; and second, lamellar, appositional growth (thickening), which proceeds on the elongating siliceous core via the formation of organic cylinders, enabling the layer-by-layer deposition of individual silica lamellae, which bio-sinter together in demosponges or remain separated, to a large extent, in hexactinellids.

2.4 Biosilica: The enzymatically formed scaffold of siliceous sponge spicules

Initially, two different mechanisms had been proposed for silicatein reaction that might describe the condensation reaction of orthosilicic acid or silicon alkoxide (TEOS) substrates (Cha et al., 1999; Fairhead et al., 2008). These models explain the hydrolytic activity of silicatein toward its synthetic substrate, TEOS, or the initial formation of disilicic acid from orthosilicic acid as the natural substrate, but none of them is able to describe the biocatalytic formation of the polymeric silica from its monomeric precursor(s).

Recently we proposed a new model, which includes the two necessary assumptions for silicatein reaction: implication of the crucial Ser moiety and using orthosilicic acid as substrate (Schröder et al., 2012a). Using this model, it is possible to explain both the initial condensation reaction, the formation of disilicic acid, and the subsequent oligomerization process. As in the previous models, the initial catalytic step involves a nucleophilic attack (S_N2 type) of the (electronegative) oxygen atom from the hydroxy group of the active site Ser to the (electropositive) silicon atom of a silicic acid molecule. It is proposed that this reaction is facilitated by hydrogen bridge formation between the Ser hydroxy and the His imidazole within the catalytic center, a reaction that increases the nucleophilicity of the Ser hydroxy group. Then, a proton transfer from the His imidazole nitrogen (Ser × His hydrogen bond) to one of the silicon OH ligands of the pentavalent intermediate (in the transition stage) occurs, resulting in the release of a water molecule. In the next step, the silicic acid molecule, which remains covalently bound through an ester-like linkage to the enzyme Ser moiety, undergoes a nucleophilic attack on the silicon atom of a second orthosilicic acid molecule, pulling it to the substrate pocket of the enzyme. Subsequently, the nucleophilicity of the attacking oxygen atom of the first (enzyme-bound) silicic acid molecule is assumed to be increased via formation of a hydrogen bond with the imidazole nitrogen of the His, in analogy to the reaction proposed (Croce et al., 2004). The loss of a water molecule generates a disilicic acid molecule, which remains bound to the enzyme. Now, a rotation of the ester bond allows the interaction of a second OH ligand of the enzyme-bound silicic acid unit with the imidazole nitrogen of the His in the catalytic center, giving rise to a further growth of the disilicic acid by a third orthosilicic acid molecule under a release of a further water molecule. Adopting previous data (Flory and Semlyen, 1966), our model proposes a final cyclization step. The products released from the enzyme by hydrolytic cleavage (cyclic trisilicic acid and higher oligomers) are much more reactive than the silicic acid monomers and promote further condensation reaction.

Enzyme kinetic studies were performed with recombinant silicatein-α and the model substrate bis(*p*-aminophenoxy)-dimethylsilane (BAPD), which contains two silicic ester-like and two silane bonds (Müller et al., 2008c). In this assay, the temperature optimum was in the range of 20–25 °C, revealing a value for the temperature coefficient (Q_{10}) that decreased by 2.5-fold above 25 °C (25 °C/35 °C) and increased by 2.9-fold below 25 °C (15 °C/25 °C). The Michaelis constant (K_m) was determined to be 22.7 µM. In comparison, the K' value for human recombinant cathepsin L, the closely related enzyme to silicatein, was 1.1 µM by using the substrate benzyloxycarbonyl-Phe-Arg-4-methylcoumarin-7-amide (Nomura et al., 1996). The turnover value (molecules of substrate converted per enzyme molecule per second) for silicatein in the silica esterase assay was 5.2, close to the value found for the human cathepsin L (Mason, 1986). Very recently, the activity of the recombinant silicatein-α was determined in *in vitro* assays and compared with the silica-forming activity during spicule formation *in vivo* (Schloßmacher et al., 2011). The silica-forming activity *in vivo* was assessed by calculating the amount of biosilica formed from the growth rate of the spicules of the demosponge *E. fluviatilis*. The growth rate of those spicules with an average size of 200–350 µm and a thickness of 15 µm was determined to be 1–10 µm/h in length (Imsiecke et al., 1995). This implies that the total weight of one spicule is approximately 88.3×10^{-9} g, which is equivalent to about 4.4×10^{13} molecules of biosilica being formed during 1 h in one spicule. Adopting a protein content of about 5% (mainly silicatein-α) in a siliceous spicule (Müller et al., 2008b), it follows that 2.2×10^{-10} g of protein catalyze the synthesis of 4.4×10^{-9} g of biosilica per h. Since the molecular mass of mature silicatein is 25 kDa, it follows that 8.8×10^{-15} mol of silicatein (5.3×10^{9} molecules) synthesize 4.4×10^{-9} g (equivalent to 7.33×10^{-11} mol [4.4×10^{13} molecules]) of biosilica per hour. In turn, and referring to native silicatein, the turnover number of silicate per silicatein on a molar basis is 8×10^{3} molecules of converted substrate per silicatein molecule per hour in spicules. With respect to recombinant silicatein-α, a value of 18.7×10^{3} molecules per molecule of silicatein-α per hour (substrate BAPD) had been determined for the esterase activity (Müller et al., 2008c) and 5.4×10^{3} molecules per molecule of silicatein-α per hour using orthosilicic acid as a substrate (Schloßmacher et al., 2011). This value can even increase by the addition of silintaphin-1 in a 4:1 molar stoichiometric ratio (silicatein-α:silintaphin-1) by greater than fivefold. Based on these calculations, it is most likely that the silica polymerization activity during spicule formation is mediated enzymatically by silicatein only (reviewed in Wang et al., 2012d).

2.5 Self-healing property of silicatein embedded in spicules

The central canal of the spicules contains the proteinaceous axial filament (Figure 2.5b) that is laterally assembled by about 10 nanofibrils (Müller et al., 2013a). By completion of the formation of the mature spicules, the axial filament remains embedded in this central structure. After treatment of the spicules with HF, the proteinaceous filaments can be recovered (Müller et al., 2013a) and stained with Coomassie brilliant blue. With progressing time, the silica is dissolved, leaving behind only the organic axial filament. It is surprising that the proteinaceous axial filament remains preserved in the spicules, even in spicules of an age of about 2 Ma, as demonstrated for the Baikalian sponge *L. baicalensis* (Müller et al., 2009a,b). The silicatein in these axial filaments remains functionally active as silica-binding protein (Cha et al., 1999),

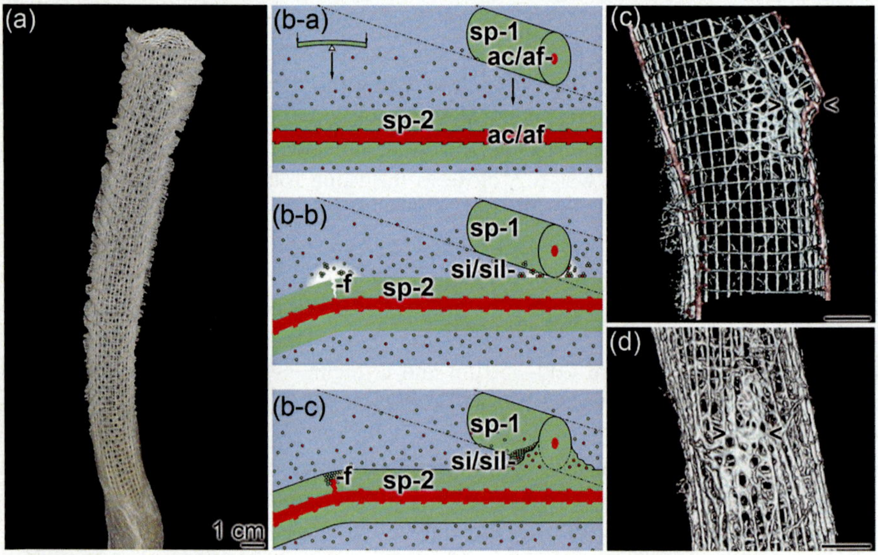

Figure 2.5 Silicatein-mediated self-healing process in sponge spicules. (a) *Euplectella aspergillum* skeleton formed of spicules that are fused to a biosilica continuum. The spicules are regularly organized in modular segments. (b) Schematic outline of the silicatein-driven self-healing process in sponge spicules. (b-a to b-c) In sponge tissue (primmorphs) that has been treated with Mn-sulfate, biosilica is un-physiologically deposited outside of two crossing spicules, spicule sp-1 and sp-2. The fusion of the two spicules occurs by biosilica deposition; biosilica is enzymatically formed from silicatein (sil) and its substrate orthosilicate (si). This fusion process is catalyzed by silicatein that is released by cells into the extracellular space; there it becomes associated with the surface of the spicules. In the second self-healing process, silicatein oozes out from the axial canal (ac), which harbors the axial filament (af). This axial filament is composed of silicatein. During this process of self-healing, silicatein is oozing out into the fissure (f) that happened in the silica mantel of the spicule. In the presence of orthosilicate, this rupture is resealed again. (c and d) Very rarely the biosilica formed spicular skeleton had been broken and reunified (> <).

or as enzyme (Schloßmacher et al., 2011), after dissolution of the biosilica shell. This functional activity of the axial filament can be demonstrated by incubating them with prehydrolyzed TEOS (orthosilicate). If the incubation is performed in the presence of 6 μg/mL of axial filament protein in a 50 mM Tris/HCl buffer (pH 7.4; 150 mM NaCl), containing 200 μM orthosilicate, bulky depositions of biosilica onto the axial filament structures can be visualized. Those structures cover almost completely the axial filaments after an incubation period of 180 min. Axial filaments do not only exist as rods, like those isolated from the tylostyles from *S. domuncula,* but also appear as radial tufts, as in asters from *G. cydonium* (Müller et al., 2007c).

After indenting and partial cracking the spicule, the silicatein molecules become exposed to the environment (Figure 2.5b) and the silica mantel crumbles. In case if orthosilicate is present in the aqueous environment at sufficiently high concentrations, the silicatein can associate with those monomers and can form biosilica. The concentrations of silica in the present-day oceans, with about 5–20 μM (Maldonado et al., 2005), might be sufficient to realize an extensive biosilica deposition as the K_m value for silicatein was determined to be 22.7 μM (Müller et al., 2008a,b,c). In turn, even spicules directly exposed to the aqueous environment have the capacity to self-heal (partially) broken spicules. If existing as skeletal elements in the body, the high concentrations of silica in the spicule-synthesizing cells, the sclerocytes, are sufficient to provide silicatein with the appropriate level of silica for the enzymatic synthesis of biosilica.

Repair, self-healing, processes taken place within spicules can be frequently recognized within hexactinellid spicules in their lamellar regions, for example, within the giant basal spicules of *M. chuni* or in the skeleton of *E. aspergillum* (Figure 2.5a–c). Fractures that have been introduced into the spicules, followed by a stay in a dry environment for a short time, remain nonsealed. However, fractures within the lamellae that occur in spicules during the growth phase fuse almost completely together (Figure 2.5c and d).

2.6 Biosilica: The osteogenic bioinorganic polymer

The formation of the skeletal system from the earliest metazoans, the sponges (phylum Porifera) (Müller et al., 2009a,b) to the crown taxa, the mammalians (Matsuo and Irie, 2008) and the insects (Moussian, 2010), is dominated by a tuned communication between cells controlling anabolic processes and cells executing catabolic reactions. Basically, two kinds of inorganic scaffold materials had been applied in the metazoan kingdom to form skeletons: calcium (calcium-based skeletal systems) and—only found in siliceous sponges—silica (silica-based skeletons).

A breakthrough in the understanding of the siliceous spicule formation of the demosponges and the hexactinellid sponges came from the discovery that the axial filaments of the spicules, the skeletal elements of these sponges (demosponges: Cha et al., 1999; hexactinellids: Müller et al., 2008a) contain an enzymatically active protein, which synthesizes polymeric silicate, the biosilica. This enzyme, silicatein, has been found to catalyze/polycondensate biosilica during the axial and radial growth of the spicules. In contrast to plant phytoliths and diatom frustules, where biosilica is deposited from a super-saturated solution onto organic templates, the siliceous spicules of sponges are formed in a hypo-saturated intra-organism environment following an enzymatic mechanism by lowering the activation energy of the polycondensation reaction. In *in vitro* systems, orthosilicic acid or TEOS (tetraethyl orthosilicate) has been used as substrate/biosilica precursor for the enzymatic reaction. During the latter process, ethanol is released (Cha et al., 1999). The silicatein-mediated formation of silica proceeds at silica substrate concentrations of around 200 μM, far below the concentrations, which are required for the chemical condensation of 1 mM or higher at neutral pH (Greenberg and Sinclair, 1955; reviewed in Wang et al., 2012a,b).

Silica is an essential nutrient both for the natural ecosystem in general (Struyf and Conley, 2009) and for humans and other vertebrates in particular (Carlisle, 1986; Van Dyck et al., 2000). Importantly, silicon deprivation results in severe skeletal malformations (Carlisle, 1972). The experimental studies showed that in avian connective tissue, the highest silicon concentrations are found, in contrast to heart or muscle tissue, where the silicon concentrations are much lower. Moreover, a spatial correlation could be established between the areas of bone formation within animal tissue and the accumulation of silicon. Based on these data, it had been concluded that a burst of silicon accumulation occurs around the osteoid and osteoid-bone interfaces, implying that this inorganic component is essential for bone formation. Consequently, we studied the effect of biosilica on the activity of osteoblasts and osteoblasts *in vitro*. SaOS-2 cells were grown in the activation cocktail on a support, coated either with hydroxyapatite (HA) or with biosilica. The cell layers that were grown on HA did not form HA crystals on their surfaces, while the cells that had been cultivated for 5 days on biosilica well formed HA crystals that are often fusing to clusters (Wiens et al., 2010a). This observation which had been supported by alizarin red S staining assays underscores that biosilica displays an inductive effect on SaOS-2 (Wiens et al., 2010a,b).

The tuned interaction of osteoblasts and osteoclasts during bone formation is well established. Furthermore, the decisive role of the osteoclastogenic ligand RANKL (ligand of the receptor activator of NF-κB) had been discussed. This ligand is processed by metalloproteases to a soluble form and interacts with RANK (receptor activator of NF-κB) and finally induces osteoclastogenesis (see Baud'huin et al., 2007). This molecular cross-talk that coordinates osteoclastogenesis is controlled by the third component, OPG (osteoprotegerin); this osteotropic effector acts as a soluble bone protector (see Feige, 2001). In turn, the molecular triad, OPG/RANK/RANKL (NF-κB ligand; see Suda et al., 1999), is not only crucially controlling osteoclast differentiation but is also involved in cell differentiation pathways of the immune and vascular systems (see Baud'huin et al., 2007). *Vice versa*, recent results provided experimental evidence demonstrating that osteoclasts contribute with their cytokines to the fine-tuning of the osteoclast/osteoblast balanced functions (see Martin and Sims, 2005).

Previous experimental evidence shows that silicate/silicon is an essential trace element in vertebrate nutrition (Carlisle, 1972; Schwarz and Milne, 1972;

Jugdaohsingh, 2007). In continuation, we showed that biosilica induces HA crystallite formation in SaOS-2 cells (Schröder et al., 2005). More recently, we demonstrated that in SaOS-2 cells, after exposure to biosilica, a differential gene expression is seen that strongly up-regulates the steady-state level of OPG transcripts and leaves the level of RANKL transcripts almost unchanged. Based on the observed increase of the OPG/RANKL ratio, it was intriguing to suggest that silica/silicate has a favorable biomedical potential for the treatment and/or prophylaxis of osteoporotic disorders (Wiens et al., 2010b; Figure 2.6). Since silica was found to elicit *in vitro*, an increased (^{3}H)dThd incorporation into DNA and a likewise increased HA formation, an osteogenic potential of silica had been deduced (Wiens et al., 2010a). Finally, we discovered that SaOS-2 cells, after exposure to biosilica, release a soluble factor, the "osteoblasts-derived inhibitory factor," which causes an inhibition of RAW 264.7 cell growth (Schröder et al., 2012b).

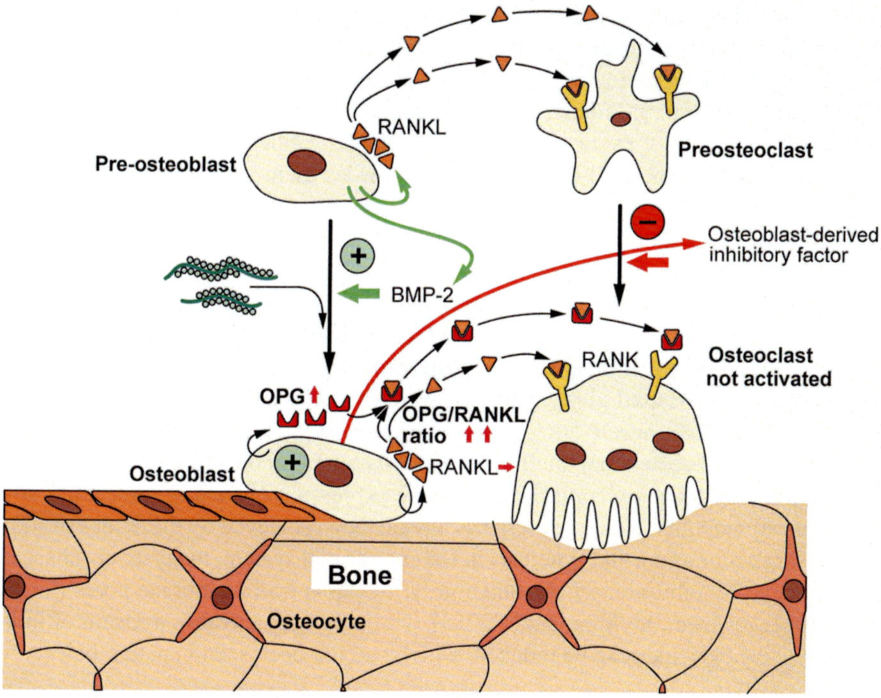

Figure 2.6 Proposed effects of biosilica on osteoblasts, osteoclasts, and their progenitor cells (schematic representation). Biosilica causes an increased expression of OPG in osteoblasts. In addition, the differentiation of osteoblasts is induced and accelerated by BMP-2 (bone morphogenetic protein 2). It is assumed that in turn the osteoblasts acquire the potential to differentiate to osteocytes and to lining cells. Furthermore, OPG counteracts various effects of RANKL, a cytokine that induces pre-osteoclast maturation and osteoclast activation. Finally, it could be identified that the osteoblasts release a factor, the osteoblasts-derived inhibitory factor, which strongly inhibits the proliferation of osteoclasts; the nature of this factor is not yet known.

2.7 Future design of novel bio-inspired, silica-based materials

The longest-lived animals, existing at present on Earth, are the species of the hexactinellid *M. chuni*, which reach an age of 11,000 years (Jochum et al., 2012). During this period, the biosilica-based giant basal spicule persisted as the main skeletal element in the bodies of this sponge. This observation reflects one of the properties of sponge biosilica to be durable and simultaneously flexible and resistant against breakage. Even more, since the major organic component of the biosilica nanohybrid particles is silicatein, it can be assumed that the biosilica material comprises a self-repair property as well (reviewed in Müller et al., 2013b). The biosilica-based sponge spicules, as intelligent structures, do not always function in the same way; they have the adaptive ability to adjust their degree of polycondensation to reach the preferred physical and chemical properties. Through the genetic and biological means provided by the silicatein-based system, the polycondensation/gelation process can be controlled and in turn the level of polymerization, coarsening and phase transformation can be adjusted. Since it is common in biological systems that reactions are reversible, a feedback control might also exist in the silicatein-based biosilica reaction, allowing a proof-reading process of the quality of the product (here the polymeric silica) formed. This property, a characteristic feature of all living systems, provides them with high flexibility and robustness. It should be mentioned that also the silicatein reaction is reversible (Müller et al., 2008c); in addition, dissolution of bioslica is catalyzed by the silicase (Schröder et al., 2003). Dynamic feedback control mechanisms are crucial elements adjusting the dynamic homeostasis of the internal environment of living systems (Bernard, 1865) and are very difficult to realize and to adapt for artificial bionic processes.

The work on biosilica in sponges already provides us with strategies to fabricate nanohybrid particle-based glass materials, which are moldable and might be used as potential biomaterials accelerating bone-repair processes (Wang et al., 2012c). In the near future, it is expected that intelligent materials based on biosilica can be produced in cell culture systems, by controlling the gene expression of the key molecules, silicatein and silintaphins. A first success has already been achieved by the demonstration that the process of aging and syneresis in the sponge cell system, the primmorphs, can be modulated by inhibition of the function of the aquaporin membrane pores (Wang et al., 2011; reviewed in Wang et al., 2013). Through this approach, a fusion of individual biosilica spicules can be induced, which had not been seen in nature before.

Acknowledgments

W.E.G. Müller is a holder of an ERC Advanced Investigator Grant (No. 268476 "BIOSILICA"). This work was supported by grants from the German Bundesministerium für Bildung und Forschung (project "Center of Excellence BIOTECmarin"), the Deutsche Forschungsgemeinschaft (SCHR 277/10-2), the European Commission (Seventh Framework Program, Marie-Curie Initial Training Network "BIOMINTEC," Grant No. 215507;

and Industry-Academia Partnerships and Pathways "CoreShell," Grant No. 286059), the International Human Frontier Science Program, the Public Welfare Project of Ministry of Land and Resources of the People's Republic of China (Grant No. 201011005–06), and the International S&T Cooperation Program of China (Grant No. 2008DFA00980).

References

Alexander, R.M., Jayes, A.S., Maloiy, G.M.O., Wathuta, E.M., 1979. Allometry of the limb bones of mammals from shrews (Sorex) to elephant (Loxodonta). J. Zool. (Lond.) 189, 305–314.

Baud'huin, M., Lamoureux, F., Duplomb, L., Rédini, F., Heymann, D., 2007. RANKL, RANK, osteoprotegerin: key partners of osteoimmunology and vascular diseases. Cell. Mol. Life Sci. 64, 2334–2350.

Bernard, C., 1865. Introduction a l'Etude de la Medecine Experimentale. Editions Garnier-Flammarion, Paris.

Biewener, A.A., 2005. Biomechanical consequences of scaling. J. Exp. Biol. 208, 1665–1676.

Carlisle, E.M., 1972. Silicon: an essential element for the chick. Science 178, 619–621.

Carlisle, E.M., 1986. Silicon as an essential trace element in animal nutrition. In: Evered, D., O'Connor, M. (Eds.), Silicon Biochemistry. CIBA Foundation Symposium, vol. 121. John Wiley & Sons Ltd, Chichester, UK, pp. 123–136.

Cha, J.N., Shimizu, K., Zhou, Y., Christianssen, S.C., Chmelka, B.F., Stucky, G.D., Morse, D.E., 1999. Silicatein filaments and subunits from a marine sponge direct the polymerization of silica and silicones *in vitro*. Proc. Natl Acad. Sci. USA 96, 361–365.

Croce, G., Frache, A., Milanesio, M., Marchese, L., Causà, M., Viterbo, D., Barbaglia, A., Bolis, V., Bavestrello, G., Cerrano, C., Benatti, U., Pozzolini, M., Giovine, M., Amenitsch, H., 2004. Structural characterization of siliceous spicules from marine sponges. Biophys. J. 86, 526–534.

Custódio, M.R., Prokic, I., Steffen, R., Koziol, C., Borojevic, R., Brümmer, F., Nickel, M., Müller, W.E.G., 1998. Primmorphs generated from dissociated cells of the sponge *Suberites domuncula*: a model system for studies of cell proliferation and cell death. Mech. Ageing Dev. 105, 45–59.

Duncan, P.M., 1881. On some remarkable enlargements of the axial canals of sponge spicules and their causes. J. R. Microsc. Soc. (Ser. 2) 1, 557–572.

Eckert, C., Schröder, H.C., Brandt, D., Perović-Ottstadt, S., Müller, W.E.G., 2006. A histochemical and electron microscopic analysis of the spiculogenesis in the demosponge *Suberites domuncula*. J. Histochem. Cytochem. 54, 1031–1040.

Fairhead, M., Johnson, K.A., Kowatz, T., Mcmahon, S.A., Carter, L.G., Oke, M., Liu, H., Naismith, J.H., van der Walle, C.F., 2008. Crystal structure and silica condensing activities of silicatein α-cathepsin L chimeras. Chem. Commun. 21, 1765–1767.

Feige, U., 2001. Osteoprotegerin. Ann. Rheum. Dis. 60, iii81–iii84.

Flory, P.J., Semlyen, J.A., 1966. Macrocyclization equilibrium constants and the statistical configuration of poly(dimethylsiloxane) chains. J. Am. Chem. Soc. 88, 3209–3212.

Greenberg, S.A., Sinclair, D., 1955. The polymerization of silicic acid. J. Phys. Chem. 59, 435–440.

Imsiecke, G., Müller, W.E.G., 1995. Unusual presence and intranuclear storage of silica crystals in the freshwater sponges *Ephydatia muelleri* and *Spongilla lacustris* (Porifera: Spongillidae). Cell. Mol. Biol. 41, 827–832.

Imsiecke, G., Steffen, R., Custodio, M., Borojevic, R., Müller, W.E.G., 1995. Formation of spicules by sclerocytes from the freshwater sponge *Ephydatia muelleri* in short-term cultures *in vitro*. In Vitro Cell Dev. Biol. Anim. 31, 528–535.

Jochum, K.P., Wang, X.H., Vennemann, T.W., Sinhal, B., Müller, W.E.G., 2012. Siliceous deep-sea sponge *Monorhaphis chuni*: a potential paleoclimate archive in ancient animals. Chem. Geol. 300–301, 143–151.

Jugdaohsingh, R., 2007. Silicon and bone health. J. Nutr. Health Aging 11, 99–110.

Maldonado, M.M., Carmona, C., Velásquez, Z., Puig, A., Cruzado, A., López, A., Young, C.M., 2005. Siliceous sponges as a silicon sink: an overlooked aspect of the benthopelagic coupling in the marine silicon cycle. Limnol. Oceanogr. 50, 799–809.

Martin, T.J., Sims, N.A., 2005. Osteoclast-derived activity in the coupling of bone formation to resorption. Trends Mol. Med. 11, 76–81.

Mason, R.W., 1986. Species variants of cathepsin L and their immunological identification. Biochem. J. 240, 285–288.

Matsuo, K., Irie, N., 2008. Osteoclast–osteoblast communication. Arch. Biochem. Biophys. 473, 201–209.

Moussian, B., 2010. Recent advances in understanding mechanisms of insect cuticle differentiation. Insect Biochem. Mol. Biol. 40, 363–375.

Müller, W.E.G., 1995. Molecular phylogeny of Metazoa [Animals]: monophyletic origin. Naturwissenschaften 82, 321–329.

Müller, W.E.G., 2001. How was the metazoan threshold crossed? The hypothetical Urmetazoa. Comp. Biochem. Physiol. 129, 433–460.

Müller, W.E.G., Rothenberger, M., Boreiko, A., Tremel, W., Reiber, A., Schröder, H.C., 2005. Formation of siliceous spicules in the marine demosponge *Suberites domuncula*. Cell Tissue Res. 321, 285–297.

Müller, W.E.G., Li, J., Schröder, H.C., Qiao, L., Wang, X.H., 2007a. The unique skeleton of siliceous sponges (Porifera; Hexactinellida and Demospongiae) that evolved first from the Urmetazoa during the Proterozoic: a review. Biogeosciences 4, 219–232.

Müller, W.E.G., Wang, X.H., Belikov, S.I., Tremel, W., Schloßmacher, U., Natoli, A., Brandt, D., Boreiko, A., Tahir, M.N., Müller, I.M., Schröder, H.C., 2007b. Formation of siliceous spicules in demosponges: example *Suberites domuncula*. In: Bäuerlein, E. (Ed.), Handbook of Biomineralization, vol. 1. The Biology of Biominerals Structure Formation. Wiley-VCH, Weinheim, pp. 59–82.

Müller, W.E.G., Schloßmacher, U., Eckert, C., Krasko, A., Boreiko, A., Ushijima, H., Wolf, S.E., Tremel, W., Müller, I.M., Schröder, H.C., 2007c. Analysis of the axial filament in spicules of the demosponge *Geodia cydonium*: different silicatein composition in microscleres [asters] and megascleres [oxeas and triaenes]. Eur. J. Cell Biol. 86, 473–487.

Müller, W.E.G., Wang, X.H., Kropf, K., Boreiko, A., Schloßmacher, U., Brandt, D., Schröder, H.C., Wiens, M., 2008a. Silicatein expression in the hexactinellid *Crateromorpha meyeri*: the lead marker gene restricted to siliceous sponges. Cell Tissue Res. 333, 339–351.

Müller, W.E.G., Jochum, K., Stoll, B., Wang, X.H., 2008b. Formation of giant spicule from quartz glass by the deep sea sponge *Monorhaphis*. Chem. Mater. 20, 4703–4711.

Müller, W.E.G., Schloßmacher, U., Wang, X.H., Boreiko, A., Brandt, D., Wolf, S.E., Tremel, W., Schröder, H.C., 2008c. Poly(silicate)-metabolizing silicatein in siliceous spicules and silicasomes of demosponges comprises dual enzymatic activities (silica-polymerase and silica-esterase). FEBS J. 275, 362–370.

Müller, W.E.G., Wang, X.H., Cui, F.Z., Jochum, K.P., Tremel, W., Bill, J., Schröder, H.C., Natalio, F., Schloßmacher, U., Wiens, M., 2009a. Sponge spicules as blueprints for the biofabrication of inorganic–organic composites and biomaterials. Appl. Microbiol. Biotechnol. 83, 397–413.

Müller, W.E.G., Schröder, H.C., Belikov, S.I., 2009b. Sustainable exploitation and conservation of the endemic Lake Baikal sponge (*Lubomirskia baicalensis*) for application in nanobiotechnology. In: Müller, W.E.G., Grachev, M.A. (Eds.), Biosilica in Evolution, Morphogenesis, and Nanobiotechnology. Progress in Molecular and Subcellular Biology [Marine Molecular Biotechnology], vol. 47. Springer, Berlin, Heidelberg, pp. 383–416.

Müller, W.E.G., Mugnaioli, E., Schröder, H.C., Schloßmacher, U., Giovine, M., Kolb, U., Wang, X.H., 2013a. Hierarchical composition of the axial filament from spicules of the siliceous sponge *Suberites domuncula*: from biosilica-synthesizing nanofibrils to structure- and morphology-guiding triangular stems. Cell Tissue Res. 351, 49–58. http://dx.doi.org/10.1007/s00441-012-1519-0.3.

Müller, W.E.G., Wang, X.H., Jochum, K.P., Schröder, H.C., 2013b. Self-healing, an intrinsic property of biomineralization processes. IUBMB Life 65, 382–396.

Nicol, D., 1966. Cope's rule and Precambrian and Cambrian invertebrates. J. Paleontol. 40, 1397–1399.

Nomura, T., Fujishima, A., Fujisawa, Y., 1996. Characterization and crystallization of recombinant human cathepsin L. Biochem. Biophys. Res. Commun. 228, 792–796.

Page, L.R., 2007. Shrinking to fit: fluid jettison from a haemocoelic hydrostatic skeleton during defensive withdrawals of a gastropod larva. Proc. Biol. Sci. 274, 2989–2994.

Schloßmacher, U., Wiens, M., Schröder, H.C., Wang, X.H., Jochum, K.P., Müller, W.E.G., 2011. Silintaphin-1: interaction with silicatein during structure guiding biosilica formation. FEBS J. 278, 1145–1155.

Schmidt-Nielsen, K., 1984. Scaling: Why Is Animal Size So Important. Cambridge University Press, Cambridge, UK.

Schröder, H.C., Krasko, A., Le Pennec, G., Adell, T., Hassanein, H., Müller, I.M., Müller, W.E.G., 2003. Silicase, an enzyme which degrades biogenous amorphous silica: contribution to the metabolism of silica deposition in the demosponge *Suberites domuncula*. In: Müller, W.E.G. (Ed.), Silicon Biomineralization: Biology-Biochemistry-Molecular Biology-Biotechnology. Prog. Mol. Subcell. Biol., vol. 33. Springer Press, Berlin, pp. 249–268.

Schröder, H.C., Borejko, A., Krasko, A., Reiber, A., Schwertner, H., Müller, W.E.G., 2005. Mineralization of SaOS-2 cells on enzymatically (Silicatein) modified bioactive osteoblast-stimulating surfaces. J. Biomed. Mat. Res. Part B Appl. Biomater. 75B, 387–392.

Schröder, H.C., Boreiko, A., Korzhev, M., Tahir, M.N., Tremel, W., Eckert, C., Ushijima, H., Müller, I.M., Müller, W.E.G., 2006. Co-expression and functional interaction of silicatein with galectin: matrix-guided formation of siliceous spicules in the marine demosponge *Suberites domuncula*. J. Biol. Chem. 281, 12001–12009.

Schröder, H.C., Wiens, M., Schloßmacher, U., Brandt, D., Müller, W.E.G., 2012a. Silicatein-mediated polycondensation of orthosilicic acid: modeling of catalytic mechanism involving ring formation. Silicon 4, 33–38.

Schröder, H.C., Wang, X.H., Wiens, M., Diehl-Seifert, B., Kropf, K., Schloßmacher, U., Müller, W.E.G., 2012b. Silicate modulates the cross-talk between osteoblasts (SaOS-2) and osteoclasts (RAW 264.7 cells): inhibition of osteoclast growth and differentiation. J. Cell. Biochem. 113, 3197–3206.

Schwarz, K., Milne, D.B., 1972. Growth-promoting effects of silicon in rats. Nature 239, 333–334.

Shimizu, K., Cha, J., Stucky, G.D., Morse, D.E., 1998. Silicatein alpha: cathepsin L-like protein in sponge biosilica. Proc. Natl Acad. Sci. USA 95, 6234–6238.

Struyf, E., Conley, D.J., 2009. Silica: an essential nutrient in wetland biogeochemistry. Front Ecol. Environ. View 7, 88–94.

Suda, T., Takahashi, N., Udagawa, N., Jimi, E., Gillespie, M.T., Martin, T.J., 1999. Modulation of osteoclast differentiation and function by the new members of the tumor necrosis factor receptor and ligand families. Endocr. Rev. 20, 345–357.

Van Dyck, K., Robberecht, H., van Cauwenbergh, R., van Vlaslaer, V., Deelstra, H., 2000. Indication of silicon essentiality in humans. Biol. Trace Elem. Res. 77, 25–32.

Walker, G., 2003. Snowball Earth: The Story of the Great Global Catastrophe That Spawned Life as We Know it. Crown, New York.

Wang, X.H., Schröder, H.C., Müller, W.E.G., 2009. Giant siliceous spicules from the deep-sea glass sponge *Monorhaphis chuni*: morphology, biochemistry, and molecular biology. Int. Rev. Cell Mol. Biol. 273, 69–115.

Wang, X.H., Schröder, H.C., Brandt, D., Wiens, M., Lieberwirth, I., Glasser, G., Schloßmacher, U., Wang, S.F., Müller, W.E.G., 2011. Sponge bio-silica formation involves syneresis following polycondensation *in vivo*. ChemBioChem 12, 2316–2324.

Wang, X.H., Schloßmacher, U., Wiens, M., Batel, R., Schröder, H.C., Müller, W.E.G., 2012a. Silicateins, silicatein interactors, and cellular interplay in sponge skeletogenesis: formation of glass fiber-like spicules. FEBS J. 279, 1721–1736.

Wang, X.H., Schröder, H.C., Wang, K., Kaandorp, J.A., Müller, W.E.G., 2012b. Genetic, biological and structural hierarchies during sponge spicule formation: from soft sol-gels to solid 3D silica composite structures. Soft Matter 5, 3657–3662.

Wang, X.H., Schröder, H.C., Wiens, M., Ushijima, H., Müller, W.E.G., 2012c. Bio-silica and bio-polyphosphate: applications in biomedicine (bone formation). Curr. Opin. Biotechnol. 21, 1–9.

Wang, X.H., Schröder, H.C., Wang, K., Kaandorp, J.A., Müller, W.E.G., 2012d. Genetic, biological and structural hierarchies during sponge spicule formation: from soft sol-gels to solid 3D silica composite structures. Soft Matter 8, 9501–9518.

Wang, X.H., Schröder, H.C., Feng, Q.L., Draenert, F., Müller, W.E.G., 2013. The deep-sea natural products, biogenic polyphosphate (Bio-PolyP) and biogenic silica (Bio-Silica), as biomimetic scaffolds for bone tissue engineering: fabrication of a morphogenetically-active polymer. Mar. Drugs 11, 718–746.

Wiens, M., Wang, X.H., Schloßmacher, U., Lieberwirth, I., Glasser, G., Ushijima, H., Schröder, H.C., Müller, W.E.G., 2010a. Osteogenic potential of bio-silica on human osteoblast-like (SaOS-2) cells. Calcif. Tissue Int. 87, 513–524.

Wiens, M., Wang, X.H., Schröder, H.C., Kolb, U., Schloßmacher, U., Ushijima, H., Müller, W.E.G., 2010b. The role of biosilica in the osteoprotegerin/RANKL ratio in human osteoblast-like cells. Biomaterials 31, 7716–7725.

Xiao, S., Kaufman, A.J., 2006. Neoproterozoic Geobiology and Paleobiology. Topics in Geobiology. Springer, Berlin.

Xiao, S., Yuan, X., Knoll, A.H., 2000. Eumetazoan fossils in terminal Proterozoic phosphorites? Proc. Natl Acad. Sci. USA 97, 13684–13689.

Part Two

Applications of marine products in tissue engineering

Bone tissue engineering using functional marine biomaterials

3

J. Venkatesan, B. Lowe, S.-K. Kim

3.1 Introduction

Tissue engineering is an emerging multidisciplinary and interdisciplinary field involving the development of bioartificial implants and/or fostering of tissue remodeling with the purpose of repairing or enhancing tissue or organ function. Bioartificial constructs generally consist of cell, biomaterials, and growth factors, so tissue engineering draws from both cell and biomaterial science and technology. Successful applications require a thorough understanding of the environment experienced by cells in normal tissues and by cells in bioartificial devices before implantation (Nerem and Sambanis, 1995). Tissue engineering seeks to provide a suitable microenvironment for cell growth and delivery of growth factors. Functionalized biomaterials can be use to fabricate three-dimensional (3D) scaffolds with appropriate pore sizes to allow cell growth, attachment, migration, and differentiation to foster tissue regeneration and/or fabrication. The 3D constructs maintain the differentiation function, architecture, and ultimately define the function of the forming bone tissue (Hutmacher, 2000). Procedures to provide the ultimate enhancement of growth factors delivery to bone grafts in the forms of autografts and allografts need to be osteoconductive and osteoinductive to enhance osteogenesis (Venkatesan et al., 2011).

Bone is a dynamic and highly vascularized tissue that continues to remodel throughout the lifetime of an individual. It plays an integral role in locomotion, ensures the skeleton has adequate load-bearing capacity, and acts as a protective casing for the delicate internal organs of the body. In addition to these structural functions, bone is intimately involved in homeostasis through its storage of Ca and P ions and by regulating the concentration of key electrolytes in the blood (Stevens, 2008). Regeneration of bone tissue is one of the most promising techniques used in orthopedic surgery and biomedical engineering by autogenous cell/tissue transplantation. The development of suitable scaffolds to serve these purpose have been under intense study and a varied number of biodegradable scaffolds and/or bioabsorbable materials have been designed and characterized at clinical and laboratory experimental levels. For a scaffold to find potential usage in the area of bone tissue regeneration, the following are important characteristics:

- 3D and highly porous with an interconnected pore network for cell growth and flow transport of nutrients and metabolic waste.
- Biocompatible and bioabsorbable with a controllable degradation and resorption rate to match cell/tissue growth *in vitro* and/or *in vivo*.
- Suitable surface chemistry for cell attachment, proliferation, and differentiation.
- Mechanical properties to match those of the tissues at the site of implantation (Hutmacher, 2000).

Functional Marine Biomaterials. http://dx.doi.org/10.1016/B978-1-78242-086-6.00003-0
Copyright © 2015 Elsevier Ltd. All rights reserved.

3.2 Bone structure

Bone is a material whose structure has been studied way back about 25,000 years ago (Weiner and Traub, 1992). The term "bone" refers to the family of materials that have their basic building blocks in the form of mineralized collagen fibril. The collagen acts as a framework in which the plate-shaped crystals of carbonated apatite forms (Weiner et al., 1999). Bone has varied arrangement of material structures that work in concert to perform diverse mechanical, biological, and chemical functions; this includes structural support, protection and storage of healing cells, and mineral ion homeostasis. Bone has various levels in hierarchical structural organizations. These levels are (1) cancellous and cortical bone; (2) the microstructure (between 10 and 500 μm); Haversian systems, osteons, single trabeculae; (3) the submicrostructure (1–10 μm); lamellae; (4) the nanostructure (from a few hundred nanometers): molecular structure of constituent elements such as fibrillar collagen and embedded mineral; and (5) the subnanostructure (below a few hundred nanometers): molecular structure of constituent elements such as mineral, collagen, and noncollagenous organic proteins. This makes bone material constituents both heterogeneous and anisotropic in nature (Rho et al., 1998).

3.3 Marine-derived biomaterials for bone tissue engineering

3.3.1 Chitosan

Chitosan is one of the widely used naturally derived polymer whose application in tissue engineering has witnessed great progress in the past two decades. It can be easily functionalized and has no toxicity. In tissue engineering applications of biomaterials, the concept of biomimetics is very important. Hence, various processes have been developed such as electrospinning, *in situ* gelling, lyophilization, and wet spinning (Di Martino et al., 2005), which are widely recognized as suitable techniques to produce ultrafine submicron and nanofibers (Zhang et al., 2008). Reports of *in vitro* studies showed that chitosan can promote growth and the deposition of much mineralized matrix in osteoblastic cell line. Chitosan sponge showed calcium deposition and alkaline phosphatase (Seol et al., 2004). A thermally induced phase separation method has been reported in the fabrication of a 3D composite for the growth of osteosarcoma MG63 cell line. Reinforced by β-tricalcium phosphate (β-TCP) and calcium phosphate, cells maintain a similar morphology on the composite and their migration could take place within a defect area (Zhang and Zhang, 2002). Osteoconductive properties of a chitosan-bonded hydroxapatite self-hardening paste have been proven by radiography showing bone tissue formation; a suitable biomaterial, which is promising for use as a bone substitute (Kawakami et al., 1992). Stimulated body fluid was used to measure the bioactivity of chitosan/nanohydroxyapatite in which the apatite layer exerted a greater degree of cell viability, proliferation, and differentiation of preosteoblast cells (Kong et al., 2006). Chitosan-alginate scaffold is very porous and has a stable compressive modulus and adds additionally to mechanical strength. Without osteogenic medium, the scaffold allowed the attachment of osteoblast and promoted the deposition of minerals *in vitro*, vascularization and tissue deposition and a calcified matrix were seen to have formed all over the scaffold (Li et al., 2005).

Hydroxyapatite (HA), which is the most stable form of calcium phosphate, is functionalized with chitosan as a composite scaffold for bone tissue engineering applications (Venkatesan and Kim, 2010). The excellent enzymatic biodegradability of chitosan and the ability for its amino and hydroxyl groups to be remodeled allows chitosan to form complexes and introduce desired functionalities for specific purposes. Platelet-derived growth factor-BB functionalized in chitosan-based scaffold allows bone formation in which osteoblastic proliferation *in vitro* was significantly enhanced. The limited usage of this scaffold for certain bone tissue regeneration purposes might be attributed to the low mechanical strength for which the compressive moduli of this scaffold, for example, were lower than the trabecular bone; such scaffold might not be recommended for use in the load-bearing bone tissue engineering applications. To cope with this problem, Jiang et al. (2006) developed microsphere-based chitosan/poly(lactic acid-glycolic acid) (PLAGA), which bears sufficient mechanical strength for use in load-bearing tissue engineering applications fostered chiefly by the PLAGA present. Preosteoblastic cells treatment (MC3T3-E1) gave bright response *in vitro*. Chitosan can play host to a number of growth factors for delivery in bone tissue regeneration; these growth factors contain bioactive agents which are capable of healing the bone, control growth and differentiation of cells and trigger angiogenesis. These growth factors include bone morphogenic proteins (BMPs), Insulin-like growth factor, and vascular endothelial growth factor. They function together during bone formation and fracture-healing processes. Some of the BMPs induce bone formation by triggering the differentiation of mesenchymal stem cells to osteoblasts and chrondroblast or perichondrial cells; this property of BMPs is prominent among the BMPs 7 and 2 and thus have been used in great number of clinical applications toward bone tissue regeneration. It is also reported that the BMP-2 factor appears as early as day 1 and BMP-7 was expressed in after 2 weeks thus their simultaneous delivery contribute much efficiently toward a viable biomimetic approach to bone healing (Yilgor et al., 2009).

The recent development of biopolymer-based nanocomposites has been used in the development of biomimetic surface attachment cues using nanomaterials for bone tissue engineering purposes. These nanomaterials have the capacity to bridge the extracellular matrix indicted response resulting from the introduction of foreign materials. Katti et al. (2008) synthesized a chitosan/montmorillonite/HA composite, which enhances cell proliferation. Similarly, chitosan-gelatin nano-bioactive glass ceramic composite has been reportedly used as a potential alveolar bone tissue engineering composite. Alveolar bone loss is a common finding associated with periodontal degeneration as many of the treatment techniques in clinic have been via filling biomaterials like bioglass to bony defects. Gelatin, which is a partial derivative of collagen, has been used as a good blend with chitosan, β-TCP, and HA for promoting cell attachment, migration, differentiation, and proliferation (Peter et al., 2010).

3.3.2 Hydroxyapatite

To better mimic the extracellular matrix as well as the mineral elements and microcomposition of natural bone, HA with polymer nano-composites widely used in the development of suitable scaffolds and forms part of a great number of sutures being used in applications for bone tissue regeneration. Natural bone is composed of inorganic part, which is largely a partially carbonated HA and a collagenous organic component. The nanoscale particle size of HA is an influential property that determines the application of HA for use in bone tissue engineering purposes. It is considered to be an essential part of bone directly responsible for its mechanical strength. Studies have also revealed that HA increases oesteoconductivity in bone. Due to its nanosize (large surface area), nano-HA is believed to have other special functions (Wei and Ma, 2004). It is largely functionalized with other marine-derived polymer such as chitosan (Zhang et al., 2008), alginate (Lin and Yeh, 2004), fucoidan (Venkatesan et al., 2014) as well as other synthetic polymers like poly(lactide-co-glycolide) (Kim et al., 2006) and poly(caprolactone) (Marra et al., 1999).

Venkatesan et al. (2011) carried out *in vitro* studies on a prepared carbon nanotube-grafted-chitosan-natural HA (isolated from *Tuna obessus*) composite. Histological analyses and morphological characterization of the scaffold showed resultant porous pore size of thermal stability, water retention, degradability, and proliferation of osteoblast cells (MG-63) in culture suitable for application in bone tissue engineering. Prerequisite for a synthetic bone substitute should provide matrix with interweaving porosity that can promote quick development of bone ingrowths, provide enough strength to prevent crushing resulting from physiological stress during integration and healing. Ability for a substitute to interact with the surrounding tissue will allow the growth of new cells. Since porous biomaterials will contribute to the regeneration of bone tissue amicably, by ensuring the attachment of cells, it is desired to use a reliable fabrication technique to achieve this property. For example, freeze casting is used to make porous complex-shaped ceramic or polymeric parts. A ceramic slurry is poured into a mold and is then frozen. This solvent will act as a binder to hold the polymeric parts together for demoulding. It is later freeze dried under vacuum; this is to prevent drying stress and shrinkage, which can cause cracks and warping during normal drying. Later, it is sintered to fabricate a porous material with improved strength, stiffness, and porosity. HA scaffold for bone tissue engineering has been reportedly fabricated using this method (Deville et al., 2006).

Yoshikawa and Myoui (2005) reported interconnected porous HA fabricated via a process called "form-jel," a technique which creates a scaffold with an excellent vascular network ample for blood supply, an aspect of bone regeneration. It serves as a good scaffold for the delivery of cytokines, which enhance bone tissue formation in defect sites. The use of poly(L-lactic acid)/HA composite scaffold for bone tissue engineering synthesized via a thermally induced phase separation technique records a high-level osteoblastic cell proliferation *in vitro* and bone-specific markers (mRNAs encoding bone sialoprotein and oesteocalcin) were also expressed significantly (Ma et al., 2001). HA scaffolds made by 3D printing are very suitable for bone replacement (Leukers et al., 2005). A standardized HA scaffold with a defined microarchitecture can be fabricated by rapid prototyping. Rapid prototyping refers to varieties of technologies capable of producing 3D physical models from 3D computer data sets. Rapid prototyping has been used largely on polymeric materials. The fabrication of HA using this method can produce porous desirable implants using fugitive molds made by stereolithography (Wilson et al., 2004). Meanwhile, a new method of preparing HA by dual-phase mixing was reported by Li et al. (2002); with a volume ratio of 1:1 resin is mixed with HA and polymethylmethacrylate (PMMA) after prolytic removal of PMMA, HA can be recovered with porous structure. This macroporous HA scaffold can be suitable for use in bone tissue engineering. The fabrication of nanohydroxyapaptite on poly(lactide-co-glycolide) acid (PLGA) and PLGA/collagen nanofibrous composite showed remarkable effect on osteoblast growth *in vitro* (Ngiam et al., 2009).

3.3.3 Collagen

The extracellular matrix is composed of a complex assemblage of structural and functional proteins, glycoproteins, and proteoglycans in a 3D construction playing a crucial role in morphogenesis, composition and function of tissues as well as organ. Collagen is a three α-polypeptide chain of one or more domains structured in a triple-helical shape of $[Gly–X–Y]_n$ composition in one of the main classes of structural ECM proteins. Collagen is biodegradable and has excellent biocompatibility. Because of its low antigenicity, hemostatic, and cell binding properties (Lee et al., 2001), collagen has become a good choice for use in bone tissue engineering. The leading industrial sources of collagen are from bone and calf skin. Chances of risks associated with bovine spongiform encephalopathy or transmissible spongiform encephalopathy are imminent. To avert this, thus, collagen can be isolated from marine organisms such as jellyfish *Stomolophus nomurai meleagris* (Song et al., 2006) and marine demosponge, *Ircinia fusca* (Pallela et al., 2011) eel fish *Evenchelys macrura* (Veeruraj et al., 2012), fish waste materials (skin, bone, and fins) (Bdolah and Keller, 1976), sea mussel *Mytilus californianus* (Bdolah and Keller, 1976), squid skin *Illex argentinus* (Kołodziejska et al., 1999), and skin of Baltic cod (*Gadus morhua*) (Sadowska et al., 2003). The fate of collagen is bone tissue engineering has been solely dictated by the above-mentioned properties. In recent times, we have seen advances in functionalizing collagen with other biopolymers or compounds in the fabrication and preparation of 3D-composite scaffolds for tissue bone tissue engineering purposes.

Using a collagen-glycosaminoglycan scaffold, Murphy et al. (2010) studied the optimal pore size required for preosteoblast adhesion, migration, and proliferation. Scaffold mean pore reported is between 85 and 325 μm and cell adhesion was greatest at 325 μm, being the mean optimal pore size for bone tissue engineering. Collagen-HA scaffold is highly osteoconductive and osteoinductive, ideal for bone tissue engineering (Rodrigues et al., 2003). Dehydrothermal treatment (DHT) used to crosslink collagen can reduce degradation rate by collagenase; this have great application potential in bone tissue engineering when collagen is functionalized in a scaffold (Wahl et al., 2007). Patterned collagen films stabilized by crosslinking carbodiimide and glutaraldehyde, DHT, and deposition of calcium phosphate on the collagen membrane was used to study the influence of pattern on cell guidance. The addition of fibrinogen induces differentiation of mesenchymal stem cells to osteoblast and cell proliferation was determined by MTT assay, osteoblastic phenotype expression was determined by alkaline phosphatase (ALP). Guidance is important when we try to mimic the 3D structure and complexity of the tissue (Ber et al., 2005; Cen et al., 2008). Wiesmann et al. (2003) revealed bone tissue engineering by primary osteoblast-like cells in a monolayer system and 3D collagen gel in which bone mineralization in the monolayer was induced by primary osteoblast. Li et al. (2006) prepared and evaluated nanohydroxyapatite/collagen/poly-L-lactic acid (PLLA) scaffold reinforced by chitin fiber for bone tissue engineering.

3.4 Recommendations and conclusion

As a result of an aging population, as well as increased injuries gotten from sports-related activities, musculoskeletal disorders have become a major healthcare problem nowadays. Current treatments although fairly successful do not give the best results. These treatments include donor tissue either from patients or other sources. The former raises great concern in terms of supply. This triggered orthopedic surgeons and research scientists to look for viable and sustainable options. In recent years, tissue engineering has gained increasing support as a method to treat orthopedic disorders. It attracts concept from principles of engineering, biology, chemistry, and materials sciences to develop suitable approaches toward the treatment of musculoskeletal disorders than using traditional methods. Meanwhile, understanding the mechanical function of bone material in relation to structure is ever captivating research, but still have challenges to solve especially as per the hierarchical organization of bone.

Tissue engineering applies bioengineering methods and is considered one of the most ideal ways by which organs that loose their functions can be regenerated and or diseases that have been too difficult to be cured. It seeks to repair injured body parts and/or organs, this can be done by using laboratory-grown tissues, implants, and materials. Technologies developed from the use of concepts gotten through research in cell signal molecules, living cells biology, and biomaterial scaffolds interaction offer greater platforms to develop regenerative systems and study their mechanisms efficiently. The marine environment is a source of varied range of biomaterials sources, which can be isolated and engineered for use in bone tissue regenerations. Efforts to foster this field of research should be strengthen, yet there are a great number of marine biomaterial resources be exploited and their functional roles determined.

Studies have shown that the demand for tissue engineering/regenerative medicine technologies has risen worldwide sharply over the past few decades. Thus, the quest to develop new technologies to respond to the growing need for implants is indeed a thing of critical importance and worth great efforts and considerations. As we work toward this as researchers, we seek to exploit the marine environment efficiently.

Acknowledgments

This research was supported by a grant from the Marine Bioprocess Research Center of the Marine Biotechnology Program, funded by the Ministry of Oceans and Fisheries, Republic of Korea.

References

Bdolah, A., Keller, P.J., 1976. Isolation of collagen granules from the foot of the sea mussel, *Mytilus californianus*. Comp. Biochem. Physiol. B 55 (2), 171–174.

Ber, S., Torun Köse, G., Hasırcı, V., 2005. Bone tissue engineering on patterned collagen films: an in vitro study. Biomaterials 26 (14), 1977–1986.

Cen, L., Liu, W., Cui, L., Zhang, W., Cao, Y., 2008. Collagen tissue engineering: development of novel biomaterials and applications. Pediatr. Res. 63 (5), 492–496.

Deville, S., Saiz, E., Tomsia, A.P., 2006. Freeze casting of hydroxyapatite scaffolds for bone tissue engineering. Biomaterials 27 (32), 5480–5489.

Di Martino, A., Sittinger, M., Risbud, M.V., 2005. Chitosan: a versatile biopolymer for orthopaedic tissue-engineering. Biomaterials 26 (30), 5983–5990.

Hutmacher, D.W., 2000. Scaffolds in tissue engineering bone and cartilage. Biomaterials 21 (24), 2529–2543.

Jiang, T., Abdel-Fattah, W.I., Laurencin, C.T., 2006. In vitro evaluation of chitosan/poly (lactic acid-glycolic acid) sintered microsphere scaffolds for bone tissue engineering. Biomaterials 27 (28), 4894–4903.

Katti, K.S., Katti, D.R., Dash, R., 2008. Synthesis and characterization of a novel chitosan/montmorillonite/hydroxyapatite nanocomposite for bone tissue engineering. Boomed. Mater. 3 (3), 034122.

Kawakami, T., Antoh, M., Hasegawa, H., Yamagishi, T., Ito, M., Eda, S., 1992. Experimental study on osteoconductive properties of a chitosan-bonded hydroxyapatite self-hardening paste. Biomaterials 13 (11), 759–763.

Kim, S.-S., Sun Park, M., Jeon, O., Yong Choi, C., Kim, B.-S., 2006. Poly(lactide-co-glycolide)/hydroxyapatite composite scaffolds for bone tissue engineering. Biomaterials 27 (8), 1399–1409.

Kołodziejska, I., Sikorski, Z.E., Niecikowska, C., 1999. Parameters affecting the isolation of collagen from squid (*Illex argentinus*) skins. Food Chem. 66 (2), 153–157.

Kong, L., Gao, Y., Lu, G., Gong, Y., Zhao, N., Zhang, X., 2006. A study on the bioactivity of chitosan/nano-hydroxyapatite composite scaffolds for bone tissue engineering. Eur. Polym. J. 42 (12), 3171–3179.

Lee, C.H., Singla, A., Lee, Y., 2001. Biomedical applications of collagen. Int. J. Pharm. 221 (1), 1–22.

Leukers, B., Gülkan, H., Irsen, S.H., Milz, S., Tille, C., Schieker, M., Seitz, H., 2005. Hydroxyapatite scaffolds for bone tissue engineering made by 3D printing. J. Mater. Sci. Mater. Med. 16 (12), 1121–1124.

Li, S.H., De Wijn, J.R., Layrolle, P., De Groot, K., 2002. Synthesis of macroporous hydroxyapatite scaffolds for bone tissue engineering. J. Biomed. Mater. Res. 61 (1), 109–120.

Li, Z., Ramay, H.R., Hauch, K.D., Xiao, D., Zhang, M., 2005. Chitosan–alginate hybrid scaffolds for bone tissue engineering. Biomaterials 26 (18), 3919–3928.

Li, X., Feng, Q., Wang, W., Cui, F., 2006. Chemical characteristics and cytocompatibility of collagen based scaffold reinforced by chitin fibers for bone tissue engineering. J. Biomed. Mater. Res. B Appl. Biomater. 77 (2), 219–226.

Lin, H.R., Yeh, Y.J., 2004. Porous alginate/hydroxyapatite composite scaffolds for bone tissue engineering: preparation, characterization, and in vitro studies. J. Biomed. Mater. Res. B Appl. Biomater. 71 (1), 52–65.

Ma, P.X., Zhang, R., Xiao, G., Franceschi, R., 2001. Engineering new bone tissue in vitro on highly porous poly(α-hydroxyl acids)/hydroxyapatite composite scaffolds. J. Biomed. Mater. Res. 54 (2), 284–293.

Marra, K.G., Szem, J.W., Kumta, P.N., DiMilla, P.A., Weiss, L.E., 1999. In vitro analysis of biodegradable polymer blend/hydroxyapatite composites for bone tissue engineering. J. Biomed. Mater. Res. 47 (3), 324–335.

Murphy, C.M., Haugh, M.G., O'Brien, F.J., 2010. The effect of mean pore size on cell attachment, proliferation and migration in collagen–glycosaminoglycan scaffolds for bone tissue engineering. Biomaterials 31 (3), 461–466.

Nerem, R.M., Sambanis, A., 1995. Tissue engineering: from biology to biological substitutes. Tissue Eng. 1 (1), 3–13.

Ngiam, M., Liao, S., Patil, A.J., Cheng, Z., Chan, C.K., Ramakrishna, S., 2009. The fabrication of nano-hydroxyapatite on PLGA and PLGA/collagen nanofibrous composite scaffolds and their effects in osteoblastic behavior for bone tissue engineering. Bone 45 (1), 4–16.

Pallela, R., Bojja, S., Janapala, V.R., 2011. Biochemical and biophysical characterization of collagens of marine sponge, *Ircinia fusca* (Porifera: Demospongiae: Irciniidae). Int. J. Biol. Macromol. 49 (1), 85–92.

Peter, M., Binulal, N., Nair, S., Selvamurugan, N., Tamura, H., Jayakumar, R., 2010. Novel biodegradable chitosan–gelatin/nano-bioactive glass ceramic composite scaffolds for alveolar bone tissue engineering. Chem. Eng. J. 158 (2), 353–361.

Rho, J.-Y., Kuhn-Spearing, L., Zioupos, P., 1998. Mechanical properties and the hierarchical structure of bone. Med. Eng. Phys. 20 (2), 92–102.

Rodrigues, C., Serricella, P., Linhares, A., Guerdes, R., Borojevic, R., Rossi, M., Duarte, M., Farina, M., 2003. Characterization of a bovine collagen–hydroxyapatite composite scaffold for bone tissue engineering. Biomaterials 24 (27), 4987–4997.

Sadowska, M., Kołodziejska, I., Niecikowska, C., 2003. Isolation of collagen from the skins of Baltic cod (*Gadus morhua*). Food Chem. 81 (2), 257–262.

Seol, Y.-J., Lee, J.-Y., Park, Y.-J., Lee, Y.-M., Rhyu, I.-C., Lee, S.-J., Han, S.-B., Chung, C.-P., 2004. Chitosan sponges as tissue engineering scaffolds for bone formation. Biotechnol. Lett. 26 (13), 1037–1041.

Song, E., Yeon Kim, S., Chun, T., Byun, H.-J., Lee, Y.M., 2006. Collagen scaffolds derived from a marine source and their biocompatibility. Biomaterials 27 (15), 2951–2961.

Stevens, M.M., 2008. Biomaterials for bone tissue engineering. Mater. Today 11 (5), 18–25.

Veeruraj, A., Arumugam, M., Ajithkumar, T., Balasubramanian, T., 2012. Isolation and characterization of drug delivering potential of type-I collagen from eel fish *Evenchelys macrura*. J. Mater. Sci. Mater. Med. 23 (7), 1729–1738.

Venkatesan, J., Kim, S.-K., 2010. Chitosan composites for bone tissue engineering – an overview. Mar. Drugs 8 (8), 2252–2266.

Venkatesan, J., Qian, Z.-J., Ryu, B., Ashok Kumar, N., Kim, S.-K., 2011. Preparation and characterization of carbon nanotube-grafted-chitosan–natural hydroxyapatite composite for bone tissue engineering. Carbohydr. Polym. 83 (2), 569–577.

Venkatesan, J., Bhatnagar, I., Kim, S.-K., 2014. Chitosan-alginate biocomposite containing fucoidan for bone tissue engineering. Mar. Drugs 12 (1), 300–316.

Wahl, D.A., Sachlos, E., Liu, C., Czernuszka, J.T., 2007. Controlling the processing of collagen-hydroxyapatite scaffolds for bone tissue engineering. J. Mater. Sci. Mater. Med. 18 (2), 201–209.

Wei, G., Ma, P.X., 2004. Structure and properties of nano-hydroxyapatite/polymer composite scaffolds for bone tissue engineering. Biomaterials 25 (19), 4749–4757.

Weiner, S., Traub, W., 1992. Bone structure: from angstroms to microns. FASEB J. 6 (3), 879–885.

Weiner, S., Traub, W., Wagner, H.D., 1999. Lamellar bone: structure–function relations. J. Struct. Biol. 126 (3), 241–255.

Wiesmann, H.P., Nazer, N., Klatt, C., Szuwart, T., Meyer, U., 2003. Bone tissue engineering by primary osteoblast-like cells in a monolayer system and 3-dimensional collagen gel. J. Oral Maxillofac. Surg. 61 (12), 1455–1462.

Wilson, C., De Bruijn, J., Van Blitterswijk, C., Verbout, A., Dhert, W., 2004. Design and fabrication of standardized hydroxyapatite scaffolds with a defined macro-architecture by rapid prototyping for bone-tissue-engineering research. J. Biomed. Mater. Res. A 68 (1), 123–132.

Yilgor, P., Tuzlakoglu, K., Reis, R.L., Hasirci, N., Hasirci, V., 2009. Incorporation of a sequential BMP-2/BMP-7 delivery system into chitosan-based scaffolds for bone tissue engineering. Biomaterials 30 (21), 3551–3559.

Yoshikawa, H., Myoui, A., 2005. Bone tissue engineering with porous hydroxyapatite ceramics. J. Artif. Organs 8 (3), 131–136.

Zhang, Y., Zhang, M., 2002. Calcium phosphate/chitosan composite scaffolds for controlled in vitro antibiotic drug release. J. Biomed. Mater. Res. 62 (3), 378–386.

Zhang, Y., Venugopal, J.R., El-Turki, A., Ramakrishna, S., Su, B., Lim, C.T., 2008. Electrospun biomimetic nanocomposite nanofibers of hydroxyapatite/chitosan for bone tissue engineering. Biomaterials 29 (32), 4314–4322.

Cardiovascular tissue engineering using functional marine biomaterials

4

E. Giuliani, V. Parravicini, R. Parravicini

4.1 Introduction

This chapter deals with the challenges of marine biomaterials for cardiovascular applications. First, it deals with structural and safety requirements inherent to their function; second, it describes the current options for human use.

Marine sources of collagen are then described with special attention to availability and possible contamination.

Finally, tuna cornea is described in detail as a source of pure collagen with peculiar antigenic characteristics that make it especially resistant to calcification.

4.2 Characteristics of cardiovascular implantable grafts

4.2.1 Structural and safety requirements

All biomaterials share a set of core characteristics that are instrumental to their application as they determine the interactions between graft and host:

- toxicity,
- biocompatibility, and
- healing.

Biomaterials, by definition, should possess an inherently low toxic potential, unless their function is to express a specific toxic activity in the site of implantation. This safety requirement is linked to the operative life of the biomaterial, which determines the release probability of toxins from the tissue.

Biocompatibility describes the interaction between host and graft: antigenicity is a key feature in this relationship as it is a major determinant of the reaction of tissues to the implanted material (Waite and Broomell, 2012). It usually leads to calcium deposition as end point of a foreign-body reaction. Calcium alters dramatically the characteristics of the material, ultimately undermining its functionality. Biocompatibility is involved not only in acute reactions but also in the long-term degradation, which affects deeply the operative life of the implant.

Another important aspect of the interaction between tissues and exogenous materials is how they influence the healing process in the vicinity of the graft, which is essential in providing a solid implant base, adequate to the required function. A poor

Functional Marine Biomaterials. http://dx.doi.org/10.1016/B978-1-78242-086-6.00004-2
Copyright © 2015 Elsevier Ltd. All rights reserved.

tissue healing around prosthesis could lead to displacement and secondary lesions with partial or complete loss of function, which in case of a cardiac valve may lead to severe consequences.

The structural and functional characteristics of cardiovascular implants vary widely according to the site and structure they have to repair or substitute, such as valves or cardiac or vascular patches. These should ideally be as similar as possible to the native counterpart in order to preserve physiology and avoid the formation of structural stress areas at the junction between tissues and biomaterials.

Finally, implants present a number of safety concerns arising from:

- The material/s they are composed of
- Antigenicity and possible allergic reactions
  - Interaction with blood and necessity of anticoagulation and/or antiplatelet therapy to ensure a correct function of the device (mechanical heart valves, synthetic vascular patches)
- Additive substances used to obtain specific characteristics (increase collagen fibers cross-linking with glutaraldehyde)
- Possible contamination from pathogens or toxic substances
  - Origin of biological compounds and possible interspecies transmission of infectious diseases
  - Quality control over breeding facilities and traceability of biomaterials

Safety is important both for synthetic and biological implants, but for the latter it translates into a constant effort to guarantee the required level of quality and control over sources and material treatment lines.

4.3 Current options, their advantages, and disadvantages

Cardiovascular implants are either synthetic or biological: as a rule of the thumb the former have a longer operative life at the expense of more complex pharmacological treatments, mainly anticoagulants and/or antiplatelet drugs, while the latter are more tolerable, requiring only a prophylactic antiplatelet treatment but suffer from a quicker degradation that affects significantly the implant durability (Thevenot et al., 2008).

Among synthetic materials, metallic compounds and polymers are the most common:

- Metals share excellent mechanical and conductivity (electrical and thermal) properties ideal for high stress applications such as heart valves. Titanium–nickel alloys have become the most common material for metallic cardiovascular applications (stents and valves) due to unique properties: shape memory effect, super-elasticity, high degree of biocompatibility; moreover, they are almost completely inert and nonmagnetic.
- Polymers show remarkable flexibility and good biocompatibility, with high resistance to biochemical attack, but their main advantage is the possibility to shape these materials as desired. Polytetrafluoroethylene is a fluorocarbon-based material often used in grafts, patches, and sutures, which can be manufactured as a porous mesh allowing cells to grow between its fibers improving biocompatibility and integration with native tissue. However, it suffers from a relatively low wear resistance that can lead to the dispersion of abrasion particles, which may trigger an inflammatory response.

Biologic materials for cardiovascular application are derived from bovine or porcine sources (Liao et al., 1992). They tend to have a better tolerability requiring less pharmacologic adjuncts such as anticoagulants therapy as in most of the cases an antiplatelet regimen is sufficient; however, they tend to degrade over time either because of reabsorption or due to calcium deposition that alters the mechanical properties of the tissue (Butany and Leask, 2001). The pace of this process may vary widely according to individual differences and site of implant.

- Bovine and porcine decellularized pericardium, treated with glutaraldehyde to increase fiber cross-linking, is the source of collagen sheets that are used either as patches or shaped as valves for cardiac valve replacement.
- Porcine complete cardiac valves are used as replacement for human valves, with good biocompatibility and functionality. Duration is their main disadvantage, leading to recurrent valvular defect and eventually to reoperation.

Animal sources of collage may raise moral concerns in patients due to religious restrictions, which exclude large parts of potential users from either form of biomaterial.

Marine sources of collagen as alternatives to terrestrial mammal-derived biomaterials and synthetic grafts:

- potential marine sources of collagen,
- edible fish as potential source of collagen and their ready availability, and
- transmissibility of pathogens to humans.

In the marine realm, several sources of collagen are available and may represent a valuable alternative to the more typically employed collagen of terrestrial mammal origin. Marine collagen may be extracted from several taxa, including jellyfish (Nagai et al., 1999), sponges (Swatschek et al., 2002), and fish (Nalinanon et al., 2007). Although some jellyfish are edible and sponges may be eventually farmed, fishes represent without doubt the marine source of collages with the largest availability on the market. Collagen may be extracted from several anatomical parts of fish body, including bones, skin, scales, and cornea. However, compared to the other fish-derived collagens, fish cornea showed several advantages. First, while the extraction of collagen from scales or bones will imply the wastage of the edible parts of fish, cornea is typically discarded by food industries and may be therefore "recycled" for other uses at low cost. Second, while the extraction of collagen from scales or bones requires complex manipulation of the original material, cornea and represents a ready-to-use material not requiring further extraction (Parravicini et al., 2012).

4.4 Tuna cornea

While the properties of cornea as a potential source of collagen for medical use are well known, its small size, especially in terrestrial mammals largely limits its employment. The Atlantic bluefin tuna (*Thunnus thynnus*) is a large pelagic predator belonging to the Scombridae family. Adult fishes are about 2 m long and may exceed 250 kg weight. Bluefin tuna dives down to a depth of about 1000 m (corresponding to about

100 bar pressure) and may reach speeds higher than 60 km/h (Froese and Pauly, 2011). The large eyes of the Atlantic bluefin tuna are devoid of eyelids and are protected by a robust cornea capable of withstanding high pressure and water friction. Being devoid of cells, the risk of pathogens which may be transmitted from tuna cornea to humans is almost null (Parravicini et al., 2012). Tuna cornea may be, however, affected by diseases induced mainly by parasites which may damage the integrity of the material. However, signs of parasites affection may be easily identified by a simple visual inspection before any further treatment (Munday et al., 2003). These characteristics make tuna cornea an ideal source of collagen for medical use.

The Atlantic bluefin tuna is highly sought, especially for sushi preparation and, consequently, thousands of tons of tuna are fished every year. The species is considered endangered and several populations of these species are being collapsing (Fromentin, 2010). Beyond any moral arguments, given the high value of the species, tuna cornea is largely available in the market. From a conservation perspective, the potential important implications of its use in medicine call for further concern regarding the protection of this species.

4.5 Tuna cornea application to cardiac valves

Tuna cornea is a pure source of collagen fiber, devoid of cellular elements: after fixation with glutaraldehyde to improve fiber cross-linking, a 40-mm disc is subdivided into three triangular portions, following collagen fibers orientation, to obtain valvular leaflets. These are then sutured onto an acetal resin support ring, which is covered with bovine pericardium patch to prevent abrasion phenomena. The prosthesis external margin is coated with polyethylene terephthalate-Dacron to facilitate suture with myocardial tissue. Hydrodynamic and durability tests showed a substantial similarity with other biologic prosthesis.

As a result, the tuna cardiac valves were tested on animal models (sheep) in mitral position. Postoperative follow-up was regular and the valves did not show signs of malfunction over a 6-month period. Histological analysis of valvular samples and X-ray and atomic absorption spectrometry demonstrated an almost absent calcium deposition.

4.6 Potential future trends

Tuna cornea is an abundant source of collagen with useful characteristics: low antigenicity and inherent safety. It can become an alternative to current available solutions for cardiac and noncardiac applications.

References

Butany, J., Leask, R., 2001. The failure modes of biological prosthetic heart valves. J. Long Term Eff. Med. Implants 11, 115–135.

Froese, R., Pauly, D. (Eds.), 2011. "*Thunnus thynnus*" in FishBase, December 2011 version.

Fromentin, J.M., 2010. The fate of Atlantic bluefin tuna. Science 327, 1325–1326.

Liao, K., Seifter, E., Hoffman, D., Yellin, E.L., Frater, R.W., 1992. Bovine pericardium versus porcine aortic valve: comparison of tissue biological properties as prosthetic valves. Artif. Organs 16, 361–365.

Munday, B.L., Sawada, Y., Cribb, T., Hayward, C.J., 2003. Diseases of tunas, *Thunnus* spp. J. Fish Dis. 26 (4), 187–206.

Nagai, T., Ogawa, T., Nakamura, T., Ito, T., Nakagawa, H., et al., 1999. Collagen of edible jellyfish exumbrella. J. Sci. Food Agric. 858, 1997–2000.

Nalinanon, S., Benjakul, S., Visessanguan, W., Kishimura, H., 2007. Use of pepsin for collagen extraction from the skin of bigeye snapper (*Priacanthus tayenus*). Food Chem. 104, 593–601. http://linkinghub.elsevier.com/retrieve/pii/S0308814606009770 (accessed 11 June 2013).

Parravicini, R., Cocconcelli, F., Verona, A., Parravicini, V., Giuliani, E., Barbieri, A., 2012. Tuna cornea as biomaterial for cardiac applications. Tex. Heart Inst. J. 39, 179–183.

Swatschek, D., Schatton, W., Kellermann, J., Müller, W.E.G., Kreuter, J., 2002. Marine sponge collagen: isolation, characterization and effects on the skin parameters surface-pH, moisture and sebum. Eur. J. Pharm. Biopharm. 53, 107–113. http://www.ncbi.nlm.nih.gov/pubmed/11777758.

Thevenot, P., Hu, W., Tang, L., 2008. Surface chemistry influences implant biocompatibility. Curr. Top. Med. Chem. 8, 270–280.

Waite, J.H., Broomell, C.C., 2012. Changing environments and structure–property relationships in marine biomaterials. J. Exp. Biol. 215, 873–883.

Skin tissue engineering using functional marine biomaterials

5

S. Colliec-Jouault

5.1 Introduction

The main strategy for skin tissue engineering is the construction of a scaffold. This scaffold is the key component for the healing process by forming a favorable environment for the cells similar to that of the extracellular matrix (ECM). The scaffold must be biocompatible, biodegradable, and must promote cell adhesion, proliferation, and organization in order to restore a functional skin. The most attractive components to create a scaffold are glycosaminoglycans (GAGs), which are naturally present in the ECM of dermis. The discovery of the biological importance of the GAGs (glycobiology), especially in the development and differentiation of cells, cell adhesion, cell signaling, and cell interactions and their ability to increase growth factor (GF) bioavailability and reduce their degradation by extracellular proteinases, has been the beginning of their crucial interest for tissue engineering. The therapeutic potential of natural bioactive compounds such as polysaccharides (especially GAGs) is now well documented, and this activity combined with their natural biodiversity will allow the development of a new generation of therapeutics. Marine environment can offer a large variety of GAG-like molecules from various origins (animals, seaweeds, invertebrates, microorganisms, etc.), which could be a good alternative to the use of mammalian GAGs as bioactive components for tissue regeneration. Recent studies have been made on newly described marine bacteria isolated from hydrothermal deep-sea vents and able to produce extracellular polysaccharides (EPS) with unusual structures. These bacterial EPS and their derivatives offer great advantages over the mammalian GAGs and other traditional marine polysaccharides.

5.2 An overview of the major advances in skin tissue engineering strategies

The development of new materials for skin tissue engineering, or new skin substitutes, requires a deep understanding of the biology of skin repair and regeneration and also their mechanisms. The moment the skin is injured (e.g., by chronic wounds such as diabetic or venous ulcers, acute burns or surgical interventions), a complex series of events begins such as inflammation, cell proliferation and migration, cell recruitment, angiogenesis, and ECM deposition to achieve both wound healing and scarring. This complex wound healing process is stimulated and controlled by various GFs and cytokines released from inflammatory cells. The crucial role of GFs for cell migration,

Functional Marine Biomaterials. http://dx.doi.org/10.1016/B978-1-78242-086-6.00005-4
Copyright © 2015 Elsevier Ltd. All rights reserved.

proliferation, and differentiation in the wounded area has been recognized (Metcalfe and Ferguson, 2007).

5.2.1 Key targets for optimum skin repair

The skin is composed of three layers, the epidermis, dermis, and hypodermis, with a complex nerve and blood supply. The epidermis is thin and highly cellular with keratinocytes and melanocytes. The dermis plays a major role in the wound healing process and biomechanics of the skin. The well-vascularized hypodermis contributes to both thermoregulatory and mechanical properties of skin. The dermis, which is the bulk of the skin, is composed of collagen, elastin, and GAGs that provide an extracellular environment (called ECM) for fibroblasts, the major cell type of dermis. Fibroblasts have a central role in the wound healing process and the ECM remodeling; wound fibroblasts not only produce remodeling enzymes such as proteases (e.g., matrix metalloproteinases) and collagenases but also can produce ECM proteins such as collagen type I and III, fibronectin and proteoglycans (PGs) with covalently attached GAG chains.

The biological importance of the GAG chains has been also recognized (Gandhi and Mancera, 2008; Mulloy, 2005). The GAG chains interact with and modulate the function of diverse proteins, including cytokines, GFs, enzymes, and adhesion molecules and thus regulate processes such as adhesion, migration, proliferation, and differentiation of cells. GAG chains and PGs are dynamic molecules with intense remodeling. Chondroitin sulfates and hyaluronic acid are the main GAGs in the skin. Chondroitin sulfate and keratan sulfate are common GAGs of the epidermis. Whereas, in the dermis, hyaluronic acid is the major component of the ECM and is also thought to have an important biological role in the wound healing. GAGs play a major role in cell signaling and development, they are believed to play a very important role in cell proliferation because they act as co-receptors for GFs, especially those from the fibroblast growth factor (FGF) family. FGFs need to interact on the cell membrane with both a heparin/heparan sulfate chain and the high affinity receptor to realize their full signaling potential (Davidson, 2008; Macri et al., 2007). GAGs play a multivalent role in the complex biochemical events of tissue repair that are hemostasis, inflammation, proliferation, and remodeling (Iozzo, 1998). GAGs, including chondroitin sulfate and hyaluronic acid, are known to act as immunosuppressants (Taylor and Gallo, 2006). During wound healing, the hyaluronic synthesis is strongly induced in both epidermis and dermis. The hyaluronan-rich matrix may contribute to a variety of cell functions that are essential for tissue repair (Juhlin, 1997). These include facilitation of cell migration into the provisional wound matrix, cell proliferation, and organization of the granulation tissue matrix. The quantity and quality of GAGs produced by fibroblasts differ in healthy skin and hypertrophic scars (Meyer et al., 2000). Consequently, not only our knowledge of the structural and functional properties of the tissue but also our understanding of the mechanisms in cell–matrix interactions and especially a better elucidation of the respective role of key biomolecules such as GFs or GAGs involved in the restoration of functional skin are essential to define a good strategy in the construct of efficient skin substitutes.

5.2.2 Skin substitutes with synthetic or natural biomaterials

Different strategies have been described to replace epidermis and dermis of injured patients. Numerous commercially epidermal or dermal constructs are available for clinical uses. Commercially *in vitro* skin substitutes are also available to perform *in vitro* experiments with the purpose of studying their efficiency in different *in vitro* models and identifying their toxicity or possible harmful side effects, as stated by the new European community regulation on chemicals and their safe use. These constructs are composed of synthetic or natural biomaterials including or not autologous or allogeneic skin cells (keratinocytes and/or fibroblasts).

5.2.2.1 Synthetic constructs

Few synthetic constructs are available because the conventional polymers are not properly designed for tissue engineering (Groeber et al., 2011). The biomaterials for tissue engineering need to be at least cytocompatible and above all biodegradable. Synthetic polymers such as polycaprolactone, polylactic acid/polyglycolic acid, polyethylene oxide, and polybutylene terephthalate are used to produce matrix, fiber, or porous substrate that are biodegradable. Some constructs are thin silicone membrane or film and also consist of a nylon matrix (Jayarama Reddy et al., 2013; Warren et al., 2003; Watanabe and Ishihara, 2003). Usually, these synthetic matrices are covered or modified to enhance their biocompatibility and are often seeded with autologous fibroblasts (dermal constructs) or keratinocytes (epidermal constructs). The major drawback of synthetic materials is the lack of cell recognition signals for engineering tissues; so often the use of natural materials that are able to mimic the nature of most tissues is more developed to design artificial skin.

5.2.2.2 Natural or biological constructs

The natural materials are often derived from skin such as, for example, collagen, gelatin, fibrin, hyaluronic acid, chondroitin sulfate and also acellular dermis from cadaveric skin origin. But materials from other origins are also widely used, such as bacterial polyesters (Valappil et al., 2007), cellulose (Teeri et al., 2007), chitosan, alginate, dextran, etc. A lot of constructs based on natural products can be found in different forms such as membrane, plate, sheet, fiber, sponge, and hydrogel.

The hydrogels are the most advantageous form because of their chemical similarity to the ECM, the rapid diffusion of hydrophilic nutrients and metabolites, their high content in water and consequently their low content in dry polymer mass (only around 0.5–20 wt%), which causes reduced irritation and much lower level of degradation products. The natural hydrogels or biopolymer-based hydrogels are degraded readily because the entire polymer backbone is exposed to water-soluble enzymes. Hydrogels can mimic the porous topography of the natural ECM and offer a three-dimensional (3D) environment that allows integration and manipulation of the cell biology of host cells. A large variety of hydrogel-based materials has been developed since 1990 (Van Vlierberghe et al., 2011). The definition often given is this of Peppas: hydrogels are water-swollen, cross-linked polymeric structure containing (1) covalent bonds produced by the reaction of one or more comonomers, (2) physical cross-links due to chain entanglements, (3) association bonds including hydrogen bonds or strong van der Waals interactions between chains, or (4) crystallites bringing together two or more macromolecular chains (Peppas et al., 2000). The hydrogel systems can be biopolymer-based using polysaccharides from different origins including the GAG family or protein-based hydrogels using mainly collagen or gelatin, but often hydrogel systems can be composed of both biopolymer and protein (Table 5.1).

Hydrogel formation can be induced by temperature or pH variation or by other stimuli such as electric or magnetic fields or ultrasound. Temperature-sensitive hydrogels are the most commonly studied class, the hydrogel constituents must be insoluble above or below a certain temperature, called lower or upper critical solution temperature (LCST or UCST, respectively). pH-sensitive hydrogels are composed of polymers with ionizable groups, known as polyelectrolytes, such as polycarboxylic acids or polyamines. The nature of functional groups affects the pH sensitivity in terms of swelling degree, depending on the basic or acidic character of the chemical groups present in the macromolecules (Elvira et al., 2002). An interesting pH-sensitive injectable cellulose-based hydrogel-bearing siloxane groups (silated hydroxypropylmethylcellulose or Si-HPMC) have been developed. The grafted cellulose-derived polymer is soluble at pH 12.8 by silane ionization into sodium silanolate ($-SiO^-Na^+$). At pH 7, the gelation of hydrogels or network formation is obtained by silanol (–SiOH) condensation due to the presence of grafted siloxane groups on Si–HPMC. This hydrogel presents a high cytocompatibility and is very convenient for the 3D culture of cells in tissue engineering (Fatimi et al., 2008; Weiss et al., 2004).

A supramolecular hydrogel comprising the phosphate-type hydrogelator 1 exhibits macroscopic gel–sol behavior in response to four distinct input stimuli: temperature, pH, Ca^{2+}, and light. This kind of supramolecular gel with multiple stimulus-response permits to elaborate gel-based supramolecular logic gates displaying AND, OR,

Table 5.1 Overview of skin substitutes with natural materials

Biopolymer-based materials
EDC cross-linked chondroitin-6-sulfate/gelatin/hyaluronan
Benzyl esters of hyaluronic acid, hyaluronan-gelatin, EDC cross-linked hyaluronan/chondroitin-6-sulfate/gelatin
Adipic dihydrazide derivatives of hyaluronic acid/PEG-propiondialdehyde, hyaluronic acid/chitosan/gelatin
DTBP-cross-linked chitosan, gold colloid/chitosan, collagen/chitosan, bFGF/chitosan
Beta-glycerol phosphate/chitosan
Gelatin/alginate
Protein-based materials
Compressed collagen, cross-linked collagen/chondroitin sulfate/hyaluronic acid, beta-glycerol phosphate/collagen/chitosan
Collagen/elastin, electrospun collagen/PCL, poly[(D,L-lactide)-co-glycolide]/collagen
Glutaraldehyde cross-linked gelatin
Collagen/elastin

NAND, and NOR functions and construct intelligent environment-sensitive supramolecular soft materials for drug delivery system (Komatsu et al., 2009). Dextran methacrylate esters have been prepared to obtain hydrogels relatively stable under physiological conditions. The possibility to adjust the cross-linking density by using dextran methacrylates with varying building units and methacrylate contents is a key factor for this approach, giving a promising way to obtain hydrogels matching the requirements of cell cultivation scaffolds (Möller et al., 2007).

From polymer solutions or gels, lyophilized wafers have been developed as drug delivery systems that can be applied directly to the surface of wounds. These wafers obtained by the freeze drying of the polymer solutions or gels can be easily sterilized with gamma irradiation. This technology offers the possibility of manufacturing sterilizable delivery systems for wounds (Matthews et al., 2006). Freeze drying is one of the most commonly used methods for the fabrication of porous biomaterials for skin tissue engineering, but often such scaffolds do not allow cells to penetrate into the inner volume because they present a layer of dense material at their surface. Cryogelation technology recently introduced can give macroporous sponge-type hydrogels; the produced scaffolds present a uniform interconnected open porous structure with a porosity of up to 90–92% and a pore size distribution of 10–20 μm. Cryogels are elastic and mechanically stable. Cryogels are produced via a cryogenic treatment of gel precursors (moderate freezing, storage in the frozen state, and subsequent thawing). Different polymers can be used to produce novel dermal regeneration templates by cryogelation, based on agarose, chondroitin sulfate, or more recently proteins such as fibrinogen and gelatin (Dainiak et al., 2010).

Beside the hydrogel strategy, collagen-based matrices are of interest for tissue repair. Collagen type I is the most abundant collagen in the body and is the most widely used substrate for the synthesis of bio-artificial skin. Highly porous three-dimensional (3D) collagen scaffolds were fabricated by a 3D dispensing system for *in vitro* culture of keratinocytes/fibroblasts for skin tissue regeneration. These porous scaffolds mimicking the ECM made from natural proteins can allow cell migration and transport of solutes through biomaterials. The collagen gel, although its poor mechanical properties, improved the graft take and decreased the delay of functional neodermis advent (Helary et al., 2005; Warren et al., 2003).

5.2.3 Skin substitutes with biological or functional components

The incorporation of cellular growth-enhancing substances or additional cell types, besides keratinocytes and fibroblasts, is also achieved to obtain constructs with improved functions and higher resemblance to native skin.

5.2.3.1 Cell-based constructs

In order to compensate for the lack of tissue and organ donors linked to potential tissue complication such as imperfect matches and dependence on immunosuppressants, the approach to create living tissue equivalents (LTEs), notably cell-seeded collagen or fibrin gels, has gained considerable attention as replacements for lost or damaged connective tissue. LTEs have several advantages over skin substitutes described above;

they allow cellularity to be achieved directly and are conducive to cell spreading and ECM formation. A LTE is traditionally made by mixing cells with a soluble biopolymer solution that sets into a gel by modifying the solution, pH, and/or temperature conditions, as described above. These LETs can be divided into epidermal equivalents or substitutes, dermal equivalents, and composite substitutes. The commercially available epidermal substitutes for clinical use are mainly composed of autologous keratinocytes and sometimes allogeneic skin cells isolated first by skin biopsy, then *in vitro* expanded prior seeding in the construct. Dermal substitutes are composed more often of neonatal allogeneic fibroblasts and sometimes of autologous fibroblasts. And the most sophisticated constructs containing both keratinocytes and fibroblasts from autologous and allogeneic skin cells are the composite substitutes with improved function and higher resemblance to native skin (Groeber et al., 2011). In the dermal substitutes, sometimes called fibroblast-populated collagen lattice, fibroblasts are cultured within a hydrated gel such as type I collagen, which undergoes cell-mediated organization and contraction to result in a tissue-like structure which, superficially at least, appears to be similar to dermis. In a recent study, the rapid production of a strong and thick natural matrix by human dermal fibroblasts was described; this cell-derived matrix is mechanically superior to LETs and could be an attractive alternative to LETs for tissue replacement (Ahlfors and Billiar, 2007).

5.2.3.2 Cytokine- and/or growth factor- and antimicrobial-based constructs

Wound healing is a process involving inflammation, cell proliferation and migration, cell recruitment, angiogenesis, formation of granulation tissue, re-epithelialization, ECM deposition and remodeling. GFs, cytokines, and chemokines released from inflammatory cells dictate the function of the cells present within wound. The presence of GFs and morphogens in constructs or in wound dressings is crucial to improve wound healing and obtain higher resemblance to native skin (Ribeiro et al., 2013). So, the incorporation of cytokines and GFs of particular importance such as members from epidermal growth factor (EGF) family or FGF family or vascular endothelial growth factor (VEGF) family or transforming growth factor-beta (TGF-beta) family into the constructs can promote the wound healing process by a better control of the release of active molecules in order to avoid harmful side effects for patient due to the exposure of high drug concentrations. These most promising GFs that require clinical testing have already been approved by the FDA and are currently used in the treatment of chronic ulcers (Barrientos et al., 2008). In the case of skin injuries, Ribeiro et al. have evaluated a dextran hydrogel loaded with chitosan microparticles containing both EGF and VEGF. In animal assays, the system elaborated by Ribeiro et al. improves the mechanical, chemical, and biological protection of the damaged skin. Moreover, the incorporation of GFs, such as VEGF and EGF, improves angiogenesis and re-epithelialization that is essential for the re-establishment of native tissue architecture (Des Rieux et al., 2011).

Antimicrobial peptides (AMPs) are currently among the promising agents for the treatments of infection. They are the main components of innate immunity and produced by various tissues and cell types of invertebrates, plants, and animals; they are peptides or basic cytokines or chemokines or neuropeptides or hormones or even defensins found in insect tissues and plant. They present several advantages over antibiotics and the most interesting is the limited capacity for the development of resistance against AMPs, so they are considered like an alternative to conventional antibiotics (Lazarev and Govorun, 2010). The same strategy described above for GFs and cytokines can be applied to AMPs, and systems can be elaborated for the controlled drug delivery of these promising therapeutic agents in order to prevent secondary infection. Antimicrobial hydrogels have been prepared through encapsulation of antimicrobial molecules into the hydrogel scaffold. Compared to conventional antibiotics, AMPs can be easily bound to polymers by chemical cross-linking, and these antibacterial hydrogels can prevent both water loss and secondary infection when exposed to the environment, absorb wound fluids and provide adequate gaseous exchange (Song et al., 2012).

5.2.3.3 GAG-based constructs

GAGs are considered like very promising macromolecules for the design of future functional biomaterials because they present major advantages over the other components. With the exception of hyaluronan, GAG chains are linked covalently to a core protein or a protein backbone and are produced by mammalian cells into the ECM in the form of PGs. These PGs form hydrated gels and fill the majority of the interstitial space; they are present in the ECM with other macromolecular components such as collagens, elastin, fibronectin, and laminin, and the GAG hyaluronan. PGs form aggregates with GAG hyaluronan as a core of the aggregate. GAG chains are large, complex carbohydrate molecules that interact with a wide range of proteins involved in physiological and pathological processes and especially with GFs. GAGs present in the ECM can bind, concentrate, and prevent the diffusion of GFs; they enhance their biological activity by improving the presentation to their relevant receptors and protect them from proteolytic degradation. Extracellular GAG chains exhibit considerable modulating effects in cell–cell and cell–ECM interactions. GAGs regulate processes such as adhesion, migration, proliferation, and differentiation of various skin cells. GAGs are hydrophilic molecules capable of absorbing up to 1000 times their volume in water to form gel-like materials. This provides dermal hydration to maintain an appropriate water balance that supports the metabolic needs of the ECM. GAG chains are unbranched polysaccharides composed of disaccharide repeating unit, which can be classified into sulfated GAGs such as heparin (HE)/heparan sulfate (HS), dermatan sulfate (DS), chondroitin sulfate (CS), and keratan sulfate (KS), and unsulfated GAGs such as hyaluronan (HA). Diverse sulfation patterns are generated *in vivo* through extensive modification of the carbohydrate backbone; GAGs present a high diversity and they are unique for each cell type. Increasing evidence suggests that GAGs encode information in the form of "sulfation code" (Gama and Hsieh-Wilson, 2005; Gandhi and Mancera, 2008; Salbach et al., 2012; Schnabelrauch et al., 2013; Van Der Smissen et al., 2011).

Different constructs have been elaborated using HA. This nonsulfated GAG plays a major role in skin wound healing, by virtue of its presence in high amounts in skin. Because of its unique physicochemical properties (e.g., viscoelastic, osmotic, hygroscopic), the nonimmunogenicity of its highly purified form and its biological properties (e.g., interactions with cells and other extracellular components), HA is a particularly attractive hydrogel material for biomedical applications and especially skin reparation. HA chains can be chemically modified or cross-linked by using different reactions to form a hydrogel in aqueous media under various conditions. The chemical modifications of HA typically involve the carboxylic acid groups and/or the alcohol groups of its backbone (Maleki et al., 2007; Milella et al., 2002; Segura et al., 2005). Current industrial production and potential biosynthetic processes of hyaluronan are already developed, leading to products of commercial significance or in development; the HA preparations are from different origins such as bacterial fermentation, rooster combs, etc. (Cimini et al., 2012; Kogan et al., 2007; Murano et al., 2011).

Heparin-based hydrogels have been also studied for their potential development as new materials for tissue engineering of skin substitutes. The development of materials or devices with better specific interactions with biological tissues is crucial. The described interactions of heparin with human skin cells suggest that it may be an excellent candidate for use in the skin delivery of drugs or cosmetics and also an active or functional component in engineered skin. Heparin was found as a good candidate for targeting skin cells and especially keratinocytes (Parisel et al., 2003). Proliferation of fibroblast cells was also observed inside a heparin gel, and a simple addition of fibrinogen during gelation was found to enhance cell proliferation inside the hydrogel more than five fold. It was possible also to add in this gel heparin-binding biomolecules such as GFs for providing additional bio-functionalities by simple adsorption. This gelation system has a potential for use as a tissue regeneration scaffold with cellular interactions via heparin and also heparin-binding factors (Tae et al., 2007). Heparin coblended with tropoelastin during the formation of synthetic elastin hydrogels increased the pore size and porosity of the hydrogels that increased colonization by dermal fibroblasts (Tu et al., 2010).

5.3 A new generation of skin substitutes with marine products

Marine natural products have a long history of exploitation, providing products for the food, personal care, cosmetic, nutraceutical, reagents, and industrial chemistry markets. Many of these products are derived from traditional industries (e.g., extraction of alginates from algae, extraction of chitin from crustacean shells). Huge opportunities remain for the exploitation of the vast biological diversity that is present in the oceans, in particular through the application of molecular biology and biochemistry technology to develop high value products. The marine biotechnology sector as a whole was estimated to be worth $2.4 billion in 2002, with a projected annual growth rate of 6–9%. The pharmaceutical market offers significant growth opportunities for marine natural products, with the potential to surpass all other sectors in terms of market value. Although the potential of Marine Biotechnology to develop new drugs is undisputed, the sector has seen limited success in the delivery of new therapeutics to the market. The approval in 2004 of Ziconotide (Prialt, Elan Pharmaceuticals), a peptide originally discovered in tropical cone snails, and in 2007 of trabectedin (Yondelis, PharmaMar) originally from a tunicate, provides important validation of marine natural products drug discovery and a number of other molecules are currently in clinical trials (Molinski et al., 2009). Marine natural products research has largely been academically inspired, but novel approaches to interdisciplinary research between marine natural products, chemists and pharmacologists, both in the academic and commercial sectors, are likely to increase the success rate in development of marine-derived therapeutics. The small but growing number of specialty marine biotechnology companies will also ensure the future success and impact of marine-derived drugs.

Marine biotechnology also has great potential for the development of novel products for the medical device and biomaterials markets. There are a number of existing marine-derived polysaccharide products such as alginate and chitosan that are used in wound dressings and biomaterials. The opportunity to develop superior hydrogels based on novel marine polysaccharides clearly exists, and the growing field of cell therapy offers enormous potential growth for the application of such products (Tombs, 1998). There is also potential for the incorporation of other functional marine biomolecules within biomaterials and medical devices to enhance their performance (e.g., immunosuppressants, hemocompatible agents). Effective collaboration between the marine biotechnologists and the medical device/biomaterial research community will ensure the delivery of new marine-derived products in this field. Traditionally, there has been a dominant focus on macro-organisms (e.g., animals, invertebrates, macroalgae), but advances in molecular techniques and microbial research are now creating significant opportunities to exploit the biological resource of the marine micro-organisms (e.g., microalgae, cyanobacteria, bacteria). Advances in analytical technology are now also creating opportunities to develop novel marine-derived biomolecules such as glycoconjugates and polysaccharides rather than small molecules, which have been the focus of much marine natural products chemistry research (Marine Board-ESF Position Paper 15, 2010).

Natural product drug discovery research in general and marine natural products in particular has been undergoing somewhat of a renaissance over recent years. The relative failure of combinatorial chemistry (synthetic)-based drug discovery has renewed the interest of big pharmaceutical companies in natural products, which historically have been the basis of about 50% of all drugs. The success of medical biotechnology in the creation of growing numbers of biopharmaceuticals has also created interest in biological molecules that previously would have been ruled out as “nondruggable.” For purposes of developing safer molecules and obtaining an alternative to molecules from mammalian origin (bovine or porcine), the production of products or materials from marine origin is a real answer to this increased demand. In fact, due to a large “species-barrier” between marine creatures and human, the probability that a prion disease could be infectious between marine creatures and human is very low. And moreover, the biomedical materials derived from marine species give benefits for people who cannot receive medical treatments using bovine- or porcine-derived medical products due to religious reasons.

5.3.1 Skin substitutes with marine products

5.3.1.1 Collagen-, gelatin-, and/or elastin-based constructs

Different studies describe the production of collagen from marine fish. A tremendous amount of fish waste has been dumped as commercial and domestic waste, which may cause pollutions and emit an offensive odor. To better utilize and reduce this waste, the extraction from fish bone, skin, or viscera of products for medical applications offers a possible way to convert this waste into a medicinally important source. Porous scaffolds composed of collagen from jellyfish were prepared by freeze drying and cross-linking to be used in tissue engineering applications. This collagen from marine origin was biostable and found to induce an immune response at least comparable to those caused by collagen from bovine origin. The jellyfish collagen exhibited higher cell viability than other naturally derived biomaterials, the collagen scaffolds presented highly porous and interconnected pore structure useful for high-density cell seeding, an efficient nutrient and an oxygen supply to cells cultured in the 3D matrices (Song et al., 2006). Type I collagen was extracted from bone of two marine fishes. This collagen was cross-linked to obtain 3D collagen films, which were efficient in animal models for healing the wounded rats covered with collagen film and promoted the granulation tissue formation (Sampath Kumar et al., 2012). Sponge containing both collagen and elastin derived from salmon were investigated as a substitute of traditional mammalian collagen sponge. Marine collagen sponge and mammalian collagen sponge were grafted onto the skin defects on the dorsum of rats. Nonsignificant differences were found between the two collagen sponges from different origins, suggesting that collagen sponges derived from marine products may be a substitute for those derived from mammals, which present a high risk of contamination by nonconventional agents such as prions or viruses (Matsumoto et al., 2011). Gelatin, an irreversibly thermally denatured and partly hydrolyzed form of collagen, can also be extracted from various marine species (e.g., shark skin, tuna head, fish skin). Few studies have focused on gelatins from invertebrate species, such as marine snail. A recent study showed that marine snail meat is a prospective source to produce gelatin in good quality with desirable functional properties, exhibiting high emulsifying stability, and water holding capacity (Zarai et al., 2012).

5.3.1.2 Chitin and chitosan-based constructs

Chitin is a natural polysaccharide and second most abundant polymer after cellulose produced annually in the world. Marine chitin is extracted from the exoskeleton of marine organisms, mainly crabs and shrimps (Laurienzo, 2010; Rinaudo, 2008). Chitin is insoluble in common solvents (such as acetone, alcohol, water), this insolubility limits its practical utilization. Chitin porous matrices can be produced from chitin solutions followed by lyophilization for cell transplantation applications to regenerate tissues. Wounds dressed with chitin films displayed improved infection-free healing rate and quality (Chow et al., 2001; Yusof et al., 2003). After solubilization, chitin can form sponges, beads, hydrogels, or films by wet spinning or freeze drying. Chitin is nontoxic and biodegradable due to chitinases widely distributed in nature (bacteria, fungi, plants, and numerous animals). Carboxymethyl chitin (CM-chitin) was one of the first derivatives prepared by chemical modification giving a water-soluble anionic polymer. Hydroxypropyl chitin derivatives were also water soluble. Chitin and chitin derivatives can be used alone or blended with other polymers.

Chitosan is the most important derivative of chitin, resulting from its deacetylation under alkaline conditions or by enzymatic hydrolysis in the presence of a chitin deacetylase. When the degree of acetylation (DA) of chitin is lower than about 50%, it becomes soluble in aqueous acidic medium and in these conditions it is named chitosan. Different chitosan derivatives can be obtained with different DA. Chitosan membranes can act as artificial skin and can be efficient skin replacement. Chitosan possesses viscosity-related properties, complete biodegradability, bacteriostatic and fungistatic properties and also bioadhesive properties (d'Ayala et al., 2008). Cell culture studies demonstrated that fibroblastic attachment and proliferation is correlated with the DA. The higher deacetylated chitosan scaffolds strongly supported the attachment and proliferation, >85% deacetylated chitosan scaffolds of 2% (W/V) composition having the highest specific growth rate and was found to be most suitable for cell culture studies and tissue engineering applications with good biostability and biocompatibility. Chitosan was found to accelerate wound healing with strong tissue-adhesive properties, chitosan 2D-shaped scaffolds such as films or porous membranes presented remarkable properties for wound dressings, whereas chitosan 3D scaffolds such as gels or sponges were really efficient as wound healing materials (Croisier and Jérôme, 2013). Research works showed that chitosan can enhance the functions of inflammatory cells such as polymorphonuclear leukocytes, macrophages, and fibroblasts promoting granulation and organization. Therefore, chitosan can be used for large open wounds with a good re-epithelialization (Malafaya et al., 2007). 3D polymeric cryogel scaffolds elaborated with natural polymers like chitosan, agarose, and gelatin allowed good cell growth and proliferation. Biochemical and microscopic analysis of the different cell types (fibroblasts and cardiac cells) showed that these scaffolds can have utility in skin and cardiac tissue engineering (Bhat and Kumar, 2012). In a very recent study, porous scaffold composite films by blending chitosan and gelatin were produced for skin tissue engineering. In a rat model, a full thickness wound was done by excising the shaved dorsal skin and the excised wound was recovered by the gelatin-chitosan film during 5 days. No sign of bacterial infection was apparent, suggesting that the films effectively protect the wounds from bacterial infection. The wounds exposed to the gelatin-chitosan film are moist and hydrated, thus demonstrating that evaporative water loss and wound dehydration are prevented. The absorption of wound exudates is clearly visible in those wounds and the re-epithelialization indicates an enhancement of the overall healing process. The scaffold with gelatin to chitosan ratio of 10:3 showed 100% healing in the rats (Rahman et al., 2012).

5.3.1.3 Alginate-based constructs

The production of commercial alginates is around 50,000 ton per year worldwide mainly extracted from algal sources, but recently some bacterial alginates are also produced extra-cellularly by *Pseudomonas aeruginosa* and *Azotobacter vinelandii*.

Thirty percentage of the production is for food industry, the rest being used in industrial, pharmaceutical, and dental applications. Alginate is one of the more abundant marine polysaccharides in pharmaceuticals. Alginate is the salt form of alginic acid; its structure is a (1–4)-linked disaccharide repeating unit composed of uronic acids: β-D-mannuronic acid (ManA or M) and α-L-guluronic acid (GulA or G). As already mentioned for the others biopolymers, both algal and bacterial alginates are biodegradable and nontoxic. Uronic acids are distributed as blocks of G or M giving poly-MG or poly-GG or poly-MM chains, M can be converted in G by epimerization using epimerases. Alginate gels are formed by either hydrogen bonding at low pH (acid gel) or ionic interaction with divalent (Ca, Cu, Zn, etc.) or trivalent (Fe, Al, etc.) ions that act as crosslinkers between adjacent polymer chains. In the controlled presence of calcium ions, alginate can gel via a process commonly known as the egg-box process, this process is also observed with pectins and carrageenans. In the case of alginate, the egg-box process involves intrachain poly-G sequences. Both swelling and gelling properties (gel strength) are dependent not only on the primary structure of alginate chains but also on their molecular weight distribution. The gel strength increases in the order: MM block < MG block < GG block so highly affected by the M:G ratio, with smaller ratios giving stronger but less flexible gels. Pure poly-M alginate chains are not found in seaweed species, but are produced by some bacteria. *Azotobacter vinelandii* like algal alginate can possess all three block sequences. However, all bacterial alginates are *O*-acetylated to varying degrees. *Azotobacter vinelandii* is the most promising bacterium in terms of industrial production because of its stable output of alginate (Coviello et al., 2007; d'Ayala et al., 2008; Laurienzo, 2010; Rinaudo, 2008; Tombs, 1998).

Wounds heal faster if kept in a moist environment with sterile exudates rather than being allowed to dry. Alginate gels have been successfully used as matrix or wound dressing because of the presence of calcium ions in exudates, alginate can form stable gel *in situ*. Calcium alginate gels, films, or fibers possess the ability to retain a moist environment and also very interesting intrinsic properties such as biocompatibility, mucoadhesion, porosity, and ease of manipulation; alginates are also good hemostatic agents. Moreover, the polymer gel can be loaded with antibiotics or wound-healing agents or mixed with other natural polymers such as chitosan or gelatin. So, alginate, alginate derivatives, or alginate blends are really interested for drug delivery, cell encapsulation, and tissue engineering. Chitosan–alginate complex was used to stimulate tissue growth through fibroblast adhesion and proliferation. Composite material based on alginate, gelatin, and wound-healing agent (cyclic adenosine monophosphate, cAMP) have been evaluated on experimental full-thickness wounds in a rat model. This *in situ*-forming hydrogels promoted wound healing leading to complete re-epithelialization of wounds within 10 days (Balakrishnan et al., 2006). A new matrix, consisting of heparin and alginate covalently crosslinked and containing bFGF, was implanted subcutaneously in the dorsal area of rats. After 2 weeks, this matrix seemed to promote cellular infiltration and angiogenesis because of its capacity to release biologically active bFGF for a long time. This matrix can stimulate the regeneration of skin by tissue engineering using an innovative artificial ECM (Tanihara et al., 2001).

5.3.2 Skin substitutes with functional marine products

Marine environment offers a rich source of original molecules, with unusual structure features, possessing naturally or by simple chemical modification, specific bioactive (or functional) properties. AMPs are small amphiphilic peptides involved in innate immune defenses that have been conserved during evolution. Invertebrates such as the molluscs and crustaceans synthesize AMPs. These molecules present a real potential as antibacterial drugs in prospective antimicrobial therapeutics; they have been evidenced also to be modulators of innate immunity. Indeed, a wide range of functions have been demonstrated for AMPs such as antiendotoxin activity, chemotaxis, promoting wound healing, and angiogenesis. AMPs can selectively modulate pro-inflammatory responses or cell proliferation and differentiation (Mookherjee and Hancock, 2007). Marine organisms offer a rich source of polysaccharides with novel structures. For example, all marine algae contain sulfated polysaccharides in their cell wall; they are a rich source of heparin-like or GAG-like entities. Sulfated polysaccharides from the three major divisions of marine algae (Rhodophyta, Phaeophyta, and Chlorophyta) have been largely studied to explore their potential as a cheap and safe source of new generation of heparinoids or GAG-like entities. Among the numerous algal polysaccharides, carrageenans from red seaweeds (Rhodophyta) and fucoidan from brown seaweeds (Phaeophyta) that share numerous properties with GAGs present a great potential to elaborate functional skin substitutes. Marine bacteria are also a source of new polysaccharides. Up to now, more than 15 bacterial strains, belonging to four main genera, have been described to produce extracellular polysaccharides or exopolysaccharides (EPS). Each EPS presents a complex and original structure that can be modified to design functional compounds and improve their specificity (Colliec-Jouault et al., 2012). Thus, the production of polysaccharides by bacteria is relevant for the development of biotechnological molecules for biomedical application, without risk for patient and environment (Rehm, 2010). The majority of microbial polysaccharides from marine source are still to be discovered.

5.3.2.1 Marine antimicrobial substance-based constructs

With evolution, marine invertebrates such as molluscs and crustaceans have developed effective systems for detecting and discriminating the micro-organisms beneficial for their physiological fitness from the ones potentially harmful and pathogenic. Given the huge biodiversity of the marine environment, one can consider that there are numerous AMPs remaining to be identified and characterized, providing opportunities characterized for development of new anti-infective agents (Bachère et al., 2004). As described above, marine AMPs could be incorporated to obtain wound-healing hydrogel with inherent antibacterial activity and additional biological activities (cell adhesion and proliferation enhancement and wound-healing acceleration).

5.3.2.2 Marine GAG-based constructs from macroalgae

Carrageenans mainly extracted from *Chondrus, Gigartina, Euchema,* and *Hypnea* are highly sulfated galactans (from 20% to 40% of sulfate groups). Carrageenans are divided into three groups with regard to differences of solubility and also the content and the position of sulfate groups on the galactose unit: iota, kappa, and lambda. The lambda-carrageenan disaccharidic repeating unit is composed of altering 2-sulfated 1,3-linked α-D-galactose and 2,6-disulfated 1,4-linked β-D-galactose (35 % of sulfate groups). Among the different groups of carrageenans, the lambda is the most hydrosoluble. Contrary to the kappa and iota, the lambda-carrageenan is a nongelling polysaccharide and presents a proportion of sulfate groups close to that found in GAGs. Low molecular weight (LMW) derivatives obtained from lambda-carrageenan share some biological properties described for GAG such as anticoagulant, antiviral, and antitumoral and can be considered like GAG mimetics (Colliec-Jouault et al., 2012).

Fucoidans, sulfated polysaccharides extracted from brown seaweeds, are among the most widely studied of all the sulfated polysaccharides of nonmammalian origin. Fucoidans are arbitrarily divided into three main families: homofucans, xylofucoglycuronans, and glycuronofucoglycans. Homofucans or fucoidans are primarily composed of L-fucopyranose units with branching or sulfate. The structures of fucoidans are heterogeneous and branched. Different linkages are described: α-(1→2) or α-(1→3) or α-(1→4) and consequently different positions of sulfate groups. A study demonstrates that the predominant repeating structure of a homogeneous fucoidan fraction from *Ascophyllum nodosum* was: [→3)-α-L-Fuc($2SO_3^-$)-(1→4)-α-L-Fuc(2,3diSO_3^-)-(1] (Chevolot et al., 2001). Fucoidans exhibit biological activities in mammalian systems as anticoagulant, antithrombotic, anti-inflammatory, antiviral, and antitumoral. These polysaccharides as heparin-like compounds and heparinoids interact with some of the heparin-binding proteins (Colliec-Jouault et al., 2012). According to their molecular weight, their chemical composition (mainly fucose and sulfate content) and their concentration, fucoidans can present contradictory effects. With two commercial preparations of fucoidan, O'Leary et al. showed that fucoidan modulates the effect of TGF-beta in *in vitro* models of dermal wound repair. A relative increase in fibroblast cell number was observed with one preparation, whereas the other preparation exhibited antiproliferative effects on dermal fibroblasts. The heterogeneity of the commercial preparations is probably the reason of these two observed opposite effects; commercial preparations contain probably different families of fucoidan with heterogeneous molecular weight (O'Leary et al., 2004). A homogeneous LMW derivative obtained from a high molecular weight (HMW) fucoidan extracted from *A. nodosum* was able *in vitro* to stimulate dermal fibroblast proliferation and accelerate collagen processing by cultured fibroblasts (Senni et al., 1998, 1999). This highly sulfated (30%) LMW fucoidan is mainly composed of fucose (40%) with a low content of uronic acid (<10%); it is a LMW fucoidan composed mainly of homofucan chains. This fucoidan can interfere on matrix metalloproteinases secretion (MMPs), factors involved in connective tissue degradation; it is able to inhibit gelatinase A (MMP-2) and stromelysin 1 (MMP-3) secretions induced by interleukin-1 beta on dermal fibroblasts in culture. Furthermore, this fucoidan increases the rate of association of MMPs with their specific inhibitors named TIMPs. Using human skin tissue sections, the ability of this fucoidan to minimize human leucocyte elastase activity was showed, resulting in the protection of human skin elastic fiber network against the enzymatic proteolysis due to this serine proteinase. This study suggests that the highly purified LMW fucoidan (rich in homofucan chains) could be used beneficially in inflammatory pathologies, such as chronic leg ulcer, emphysema, or periodontitis in which uncontrolled ECM degradation occurs (Senni et al., 2006a). Other studies on this highly purified LMW fucoidan showed that it can bind and modulate the activity of proangiogenic GFs such as FGF and VEGF, can also promote tissue rebuilding parameters such as signaling by heparin-binding GFs (FGF-2 and VEGF) and collagen processing in fibroblasts (Matou et al., 2002). Some anti-inflammatory properties such as anticomplementary activities with both inhibition of leucocyte margination and connective tissue proteolysis have been demonstrated for this LMW fucoidan (Senni et al., 2011).

5.3.2.3 Marine GAG-based constructs from marine bacteria

Bacteria can synthesize a wide range of biopolymers, especially extracellular polysaccharides such as EPS, endowed with diverse biological functions suitable for numerous industrial and medical applications. Bacterial EPS can be produced at a viable economic cost, completely in controlled conditions in agreement with the Good Manufacturing Practices (GMPs) (axenic culture, well-defined medium culture, sterilized conditions, etc.) through large-scale fermentations. Dextran and xanthan have annual world production volumes of around 2000 and 100,000 ton, respectively. Bacterial EPS production is greatly influenced by fermentation conditions; so, in modifying several factors (such as the composition of medium culture mainly carbon and nitrogen sources, pH, temperature, agitation, or oxygen concentration), the structure, composition, and viscosity of EPS can be modified. Moreover, a good understanding of the biosynthesis of the EPS and its regulation can allow to produce tailor-made polymers (Nicolaus et al., 2010; Rehm, 2010; Tombs, 1998).

The bacterial EPS from terrestrial origin presents a great structural diversity; but this diversity is tremendously increased in the marine environment due to the enormity of the marine biosphere. In fact, marine environment is an extraordinary untapped source of unusual bacteria that remain to be discovered and offer a rich potential for both biotechnology and drug discovery. Major recent advances in deep-sea biotechnology have allowed to isolate novel micro-organisms (Deming, 1998; Guezennec, 2002). A very interesting new EPS-producing *Vibrio* has been discovered from deep-sea environment (Raguenes et al., 1997a). It is the first species of *Vibrio* to be isolated from a vent sample and it can produce a very innovative EPS presenting a chemical resemblance to HA and HS after chemical modification. The structure of this EPS, named HE800 EPS, consists of a linear tetrasaccharide repeating unit: two glucuronic acid, one *N*-acetylated glucosamine, and one *N*-acetylated galactosamine residues. It is very rare to find hexosamine sugars in bacterial EPS; it is a HMW EPS ($>10^6$ g/mol) (Rougeaux et al., 1999). This unusual HE800 EPS with a GAG-like structure presents also GAG-like biological properties. HE800 EPS is a good candidate to construct skin substitutes or dermal equivalents with functional properties especially in association with fibrillar collagens (Figure 5.1). It presents HA-like activities; it can promote both collagen structuring in dermal equivalent and fibroblast colonization of this reconstructed tissue. HE800 EPS derivatives were produced by depolymerization and substitution reactions in order to obtain derivatives having a

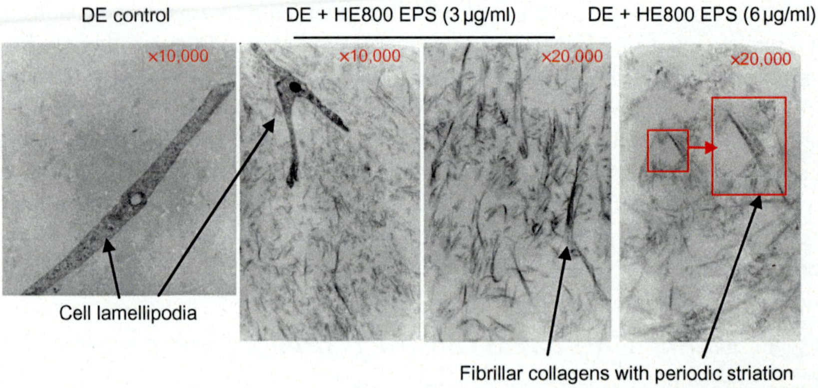

Figure 5.1 Dermal equivalent (DE) performed with collagen I (600 μg/ml) and HMW HE800 EPS (0, 3, or 6 μg/ml). Human normal dermal fibroblasts (3×10^5 cells/ml) were embedded within gel before collagen polymerization. After 11 days of culture, reconstructed tissues were fixed and prepared for electron microscopy or histology and image analysis studies.

molecular weight <30 kg/mol and a sulfate content of 30%. These derivatives can be considered as a new class of GAG-like derivatives exhibiting HS properties. HE800 EPS derivatives were potent inhibitors of the complement cascade and 10 times less anticoagulant than heparin in clotting assays. They improved the dermal fibroblast proliferation in 2D cultures and can inhibit IL-1beta stimulation of MMPs secretion. To summarize, these derivatives are able to inhibit some parameters involved in tissue breakdown and inflammation such as complement cascade and induction of MMPS by IL-1beta. They are also able to induce tissue repair in promoting GFs activity and consequently fibroblast proliferation. HE800 EPS and its derivatives produced by *Vibrio diabolicus* can be considered potentially as new functional biomaterial for tissue repair and engineering (Senni et al., 2006b, 2013).

Another new bacterium was also isolated in the vicinity of an active hydrothermal vent of the southern depression of the Guaymas basin (Gulf of California); it is a deep-sea aerobic mesophilic heterotrophic bacterium described as a new species of the genus *Alteromonas* and the proposed name is *Alteromonas infernus* (Raguenes et al., 1997b). This bacterium *A. infernus* can produce a water-soluble EPS, named GY785 EPS; it is a branched heteropolysaccharide and its repeating unit is a nonasaccharide: four glucose residues, two galactose, two glucuronic acids, and one galacturonic acid bearing a sulfate group at C_2 position (Roger et al., 2004). As described above for HE800 EPS, GY785 EPS derivatives were isolated. These derivatives had a lower anticoagulant activity than heparin (Colliec Jouault et al., 2001), promoted *in vitro* angiogenesis by its positive effects on vascular tube formation and cell proliferation via major angiogenic GFs such as FGF-2 and VEGF (Matou et al., 2005). Added in a chondrogenic differentiation medium, comprising chondrogenic GFs such as TGF-β1, GY785 EPS derivatives can enhance the chondrogenic differentiation of mesenchymal stem cells, considered as an attractive source of cells for cartilage engineering (Merceron et al., 2012). It was also demonstrated that GY785

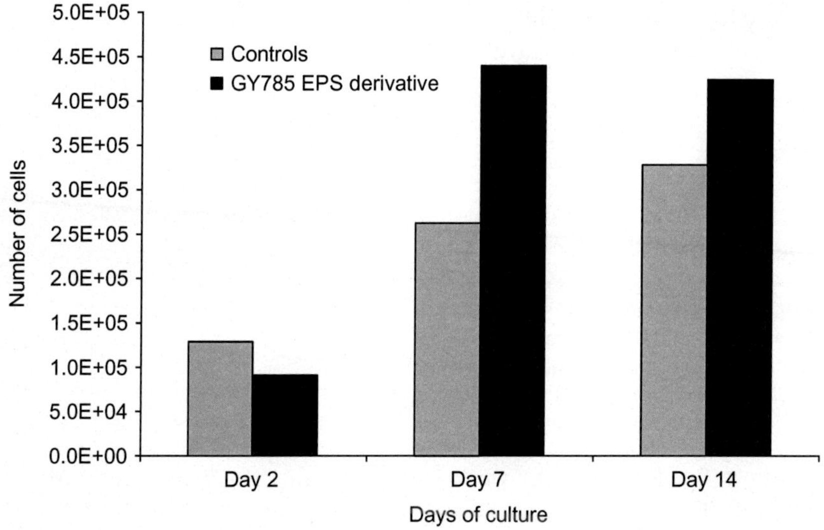

Figure 5.2 Dermal equivalent (DE) performed with collagen I (600 μg/ml) and GY785 EPS (10 μg/ml). Human normal dermal fibroblasts (3×10^5 cells/ml) were embedded within gel before collagen polymerization.

EPS derivatives stimulated some human mesenchymal cells such as dermal fibroblasts, gingival fibroblasts, and stromal medullar cells. In 2D cultures, GY785 EPS derivatives promoted FGF-2 signaling and thus cell proliferation; this effect was also observed when cells were mixed with extracellular collagenous matrix (Figure 5.2). As described for HE800 EPS derivatives, GY785 EPS derivatives can inhibit some processes involved in tissue breakdown and inflammation such as induction of MMPs by inflammatory cytokines (IL-1beta and TNF-alpha) and complement cascade (Senni et al., 2005, 2011). The marine bacterial EPS could provide biochemical entities with suitable functions for obtaining effective compounds for tissue engineering.

5.4 Conclusion

Among the numerous strategies developed to isolate new materials for obtaining efficient innovate skin substitutes, the use of marine molecules is of a high interest for the future and will contribute undoubtedly to answer to the grand challenges for optimal skin repair. Functional marine GAG-like molecules present a real potential to obtain efficient skin-substitutes inducing tissue repair by improving GF signaling such as FGF-2 and VEGF. Marine bacterial GAG-like molecules (e.g., EPS and EPS derivatives) have especially some great advantages because they can be produced at viable economic cost, in controlled conditions in agreement with the GMPs and with a very low risk for patients to be infected by a nonconventional transmissible agent such as prions or emerging viruses due to a large "species-barrier." These biotechnological marine GAG-like molecules and their derivatives present often very original structures and have a better benefit/risk ratio than GAGs from mammalian origins (e.g., heparin).

References

Ahlfors, J.E.W., Billiar, K.L., 2007. Biomechanical and biochemical characteristics of a human fibroblast-produced and remodeled matrix. Biomaterials 28, 2183–2191.

Bachère, E., Gueguen, Y., Gonzalez, M., De Lorgeril, J., Garnier, J., Romestand, B., 2004. Insights into the anti-microbial defense of marine invertebrates: the penaeid shrimps and the oyster *Crassostrea gigas*. Immunol. Rev. 198, 149–168.

Balakrishnan, B., Mohanty, M., Fernandez, A.C., Mohanan, P.V., Jayakrishnan, A., 2006. Evaluation of the effect of incorporation of dibutyryl cyclic adenosine monophosphate in an *in situ*-forming hydrogel wound dressing based on oxidized alginate and gelatin. Biomaterials 27, 1355–1361.

Barrientos, S., Stojadinovic, O., Golinko, M.S., Brem, H., Tomic-Canic, M., 2008. PERSPECTIVE ARTICLE: growth factors and cytokines in wound healing. Wound Repair Regen. 16, 585–601.

Bhat, S., Kumar, A., 2012. Cell proliferation on three-dimensional chitosan-agarose-gelatin cryogel scaffolds for tissue engineering applications. J. Biosci. Bioeng. 114, 663–670.

Chevolot, L., Mulloy, B., Ratiskol, J., Foucault, A., Colliec-Jouault, S., 2001. A disaccharide repeat unit is the major structure in fucoidans from two species of brown algae. Carbohydr. Res. 330, 529–535.

Chow, K.S., Khor, E., Wan, A.C.A., 2001. Porous chitin matrices for tissue engineering: fabrication and *in vitro* cytotoxic assessment. J. Polym. Res. 8, 27–35.

Cimini, D., De Rosa, M., Schiraldi, C., 2012. Production of glucuronic acid-based polysaccharides by microbial fermentation for biomedical applications. Biotechnol. J. 7, 237–250.

Colliec-Jouault, S., Bavington, C., Delbarre-Ladrat, C., 2012. Heparin-like entities from marine organisms. Handb. Exp. Pharmacol. 207, 423–449.

Colliec-Jouault, S., Chevolot, L., Helley, D., Ratiskol, J., Bros, A., Sinquin, C., Roger, O., Fischer, A.M., 2001. Characterization, chemical modifications and *in vitro* anticoagulant properties of an exopolysaccharide produced by *Alteromonas infernus*. Biochim. Biophys. Acta 1528, 141–151.

Coviello, T., Matricardi, P., Marianecci, C., Alhaique, F., 2007. Polysaccharide hydrogels for modified release formulations. J. Control. Release 119, 5–24.

Croisier, F., Jérôme, C., 2013. Chitosan-based biomaterials for tissue engineering. Eur. Polym. J. 49, 780–792.

Dainiak, M.B., Allan, I.U., Savina, I.N., Cornelio, L., James, E.S., James, S.L., Mikhalovsky, S.V., Jungvid, H., Galaev, I.Y., 2010. Gelatin-fibrinogen cryogel dermal matrices for wound repair: preparation, optimisation and *in vitro* study. Biomaterials 31, 67–76.

d'Ayala, G.G., Malinconico, M., Laurienzo, P., 2008. Marine derived polysaccharides for biomedical applications: chemical modification approaches. Molecules 13, 2069–2106.

Davidson, J.M., 2008. First-class delivery: getting growth factors to their destination. J. Invest. Dermatol. 128, 1360–1362.

Deming, J., 1998. Deep ocean environmental biotechnology. Curr. Opin. Biotechnol. 9, 283–287.

des Rieux, A., Ucakar, B., Mupendwa, B.P.K., Colau, D., Feron, O., Carmeliet, P., Préat, V., 2011. 3D systems delivering VEGF to promote angiogenesis for tissue engineering. J. Control. Release 150, 272–278.

Elvira, C., Mano, J.F., San Román, J., Reis, R.L., 2002. Starch-based biodegradable hydrogels with potential biomedical applications as drug delivery systems. Biomaterials 23, 1955–1966.

Fatimi, A., Tassin, J.F., Quillard, S., Axelos, M.A.V., Weiss, P., 2008. The rheological properties of silated hydroxypropylmethylcellulose tissue engineering matrices. Biomaterials 29, 533–543.

Gama, C.I., Hsieh-Wilson, L.C., 2005. Chemical approaches to deciphering the glycosaminoglycan code. Curr. Opin. Chem. Biol. 9, 609–619.

Gandhi, N.S., Mancera, R.L., 2008. The structure of glycosaminoglycans and their interactions with proteins. Chem. Biol. Drug Des. 72, 455–482.

Groeber, F., Holeiter, M., Hampel, M., Hinderer, S., Schenke-Layland, K., 2011. Skin tissue engineering – *In vivo* and *in vitro* applications. Adv. Drug Del. Rev. 63, 352–366.

Guezennec, J., 2002. Deep-sea hydrothermal vents: a new source of innovative bacterial exopolysaccharides of biotechnological interest? J. Ind. Microbiol. Biotechnol. 29, 204–208.

Helary, C., Foucault-Bertaud, A., Godeau, G., Coulomb, B., Giraud Guille, M.M., 2005. Fibroblast populated dense collagen matrices: cell migration, cell density and metalloproteinases expression. Biomaterials 26, 1533–1543.

Iozzo, R.V., 1998. Matrix proteoglycans: from molecular design to cellular function. Annu. Rev. Biochem. 67, 609–652.

Jayarama Reddy, V., Radhakrishnan, S., Ravichandran, R., Mukherjee, S., Balamurugan, R., Sundarrajan, S., Ramakrishna, S., 2013. Nanofibrous structured biomimetic strategies for skin tissue regeneration. Wound Repair Regen. 21, 1–16.

Juhlin, L., 1997. Hyaluronan in skin. J. Intern. Med. 242, 61–66.

Kogan, G., Šoltés, L., Stern, R., Gemeiner, P., 2007. Hyaluronic acid: a natural biopolymer with a broad range of biomedical and industrial applications. Biotechnol. Lett. 29, 17–25.

Komatsu, H., Matsumoto, S., Tamaru, S.-I., Kaneko, K., Ikeda, M., Hamachi, I., 2009. Supramolecular hydrogel exhibiting four basic logic gate functions to fine-tune substance release. J. Am. Chem. Soc. 131, 5580–5585.

Laurienzo, P., 2010. Marine polysaccharides in pharmaceutical applications: an overview. Mar. Drugs 8, 2435–2465.

Lazarev, V.N., Govorun, V.M., 2010. Antimicrobial peptides and their use in medicine. Appl. Biochem. Microbiol. 46, 803–814.

Macri, L., Silverstein, D., Clark, R.A.F., 2007. Growth factor binding to the pericellular matrix and its importance in tissue engineering. Adv. Drug Del. Rev. 59, 1366–1381.

Malafaya, P.B., Silva, G.A., Reis, R.L., 2007. Natural-origin polymers as carriers and scaffolds for biomolecules and cell delivery in tissue engineering applications. Adv. Drug Del. Rev. 59, 207–233.

Maleki, A., Kjoniksen, A.-L., Nystrom, B., 2007. Characterization of the chemical degradation of hyaluronic acid during chemical gelation in the presence of different cross-linker agents. Carbohydr. Res. 342, 2776–2792.

Marine Board-ESF Position, Paper 15, September 2010. Marine Biotechnology: A New Vision and Strategy for Europe. http://www.esf.org/publications/marine-sciences.html.

Matou, S., Helley, D., Chabut, D., Bros, A., Fischer, A.-M., 2002. Effect of fucoidan on fibroblast growth factor-2-induced angiogenesis *in vitro*. Thromb. Res. 106, 213–221.

Matou, S., Colliec-Jouault, S., Galy-Fauroux, I., Ratiskol, J., Sinquin, C., Guezennec, J., Fischer, A.-M., Helley, D., 2005. Effect of an oversulfated exopolysaccharide on angiogenesis induced by fibroblast growth factor-2 or vascular endothelial growth factor *in vitro*. Biochem. Pharmacol. 69, 751–759.

Matsumoto, Y., Ikeda, K., Yamaya, Y., Yamashita, K., Saito, T., Hoshino, Y., Koga, T., Enari, H., Suto, S., Yotsuyanagi, T., 2011. The usefulness of the collagen and elastin sponge derived from salmon as an artificial dermis and scaffold for tissue engineering. Biomed. Res. 32, 29–36.

Matthews, K.H., Stevens, H.N.E., Auffret, A.D., Humphrey, M.J., Eccleston, G.M., 2006. Gamma-irradiation of lyophilised wound healing wafers. Int. J. Pharm. 313, 78–86.

Merceron, C., Portron, S., Vignes-Colombeix, C., Rederstorff, E., Masson, M., Lesoeur, J., Sourice, S., Sinquin, C., Colliec-Jouault, S., Weiss, P., Vinatier, C., Guicheux, J., 2012. Pharmacological modulation of human mesenchymal stem cell chondrogenesis by a chemically oversulfated polysaccharide of marine origin: potential application to cartilage regenerative medicine. Stem cells (Dayton, Ohio) 30, 471–480.

Metcalfe, A.D., Ferguson, M.W.J., 2007. Bioengineering skin using mechanisms of regeneration and repair. Biomaterials 28, 5100–5113.

Meyer, L.J.M., Russell, S.B., Russell, J.D., Trupin, J.S., Egbert, B.M., Shuster, S., Stern, R., 2000. Reduced hyaluronan in keloid tissue and cultured keloid fibroblasts. J. Invest. Dermatol. 114, 953–959.

Milella, E., Brescia, E., Massaro, C., Ramires, P.A., Miglietta, M.R., Fiori, V., Aversa, P., 2002. Physico-chemical properties and degradability of non-woven hyaluronan benzylic esters as tissue engineering scaffolds. Biomaterials 23, 1053–1063.

Molinski, T.F., Dalisay, D.S., Lievens, S.L., Saludes, J.P., 2009. Drug development from marine natural products. Nat. Rev. Drug Discov. 8, 69–85.

Möller, S., Weisser, J., Bischoff, S., Schnabelrauch, M., 2007. Dextran and hyaluronan methacrylate based hydrogels as matrices for soft tissue reconstruction. Biomol. Eng. 24, 496–504.

Mookherjee, N., Hancock, R.E.W., 2007. Cationic host defence peptides: innate immune regulatory peptides as a novel approach for treating infections. Cell. Mol. Life Sci. 64, 922–933.

Mulloy, B., 2005. The specificity of interactions between proteins and sulfated polysaccharides. An. Acad. Bras. Cienc. 77, 651–664.

Murano, E., Perin, D., Khan, R., Bergamin, M., 2011. Hyaluronan: from biomimetic to industrial business strategy. Nat. Prod. Commun. 6, 555–572.

Nicolaus, B., Kambourova, M., Oner, E.T., 2010. Exopolysaccharides from extremophiles: from fundamentals to biotechnology. Environ. Technol. 31, 1145–1158.

O'Leary, R., Rerek, M., Wood, E.J., 2004. Fucoidan modulates the effect of transforming growth factor (TGF)-beta1 on fibroblast proliferation and wound repopulation in *in vitro* models of dermal wound repair. Biol. Pharmaceut. Bull. 27, 266–270.

Parisel, C., Saffar, L., Gattegno, L., André, V., Abdul-Malak, N., Perrier, E., Letourneur, D., 2003. Interactions of heparin with human skin cells: binding, location, and transdermal penetration. J. Biomed. Mater. Res. A 67A, 517–523.

Peppas, N.A., Bures, P., Leobandung, W., Ichikawa, H., 2000. Hydrogels in pharmaceutical formulations. Eur. J. Pharm. Biopharm. 50, 27–46.

Raguenes, G., Christen, R., Guezennec, J., Pignet, P., Barbier, G., 1997a. *Vibrio diabolicus* sp. nov., a new polysaccharide-secreting organism isolated from a deep-sea hydrothermal vent polychaete annelid, *Alvinella pompejana*. Int. J. Syst. Bacteriol. 47, 989–995.

Raguenes, G.H., Peres, A., Ruimy, R., Pignet, P., Christen, R., Loaec, M., Rougeaux, H., Barbier, G., Guezennec, J.G., 1997b. *Alteromonas infernus* sp. nov., a new polysaccharide-producing bacterium isolated from a deep-sea hydrothermal vent. J. Appl. Microbiol. 82, 422–430.

Rahman, M.M., Pervez, S., Nesa, B., Khan, M.A., 2012. Preparation and characterization of porous scaffold composite films by blending chitosan and gelatin solutions for skin tissue engineering. Polym. Int. 62, 79–86.

Rehm, B.H.A., 2010. Bacterial polymers: biosynthesis, modifications and applications. Nat. Rev. Microbiol. 8, 578–592.

Ribeiro, M.P., Morgado, P.I., Miguel, S.P., Coutinho, P., Correia, I.J., 2013. Dextran-based hydrogel containing chitosan microparticles loaded with growth factors to be used in wound healing. Mater. Sci. Eng. C 33, 2958–2966.

Rinaudo, M., 2008. Main properties and current applications of some polysaccharides as biomaterials. Polym. Int. 57, 397–430.

Roger, O., Kervarec, N., Ratiskol, J., Colliec-Jouault, S., Chevolot, L., 2004. Structural studies of the main exopolysaccharide produced by the deep-sea bacterium *Alteromonas infernus*. Carbohydr. Res. 339, 2371–2380.

Rougeaux, H., Kervarec, N., Pichon, R., Guezennec, J., 1999. Structure of the exopolysaccharide of *Vibrio diabolicus* isolated from a deep-sea hydrothermal vent. Carbohydr. Res. 322, 40–45.

Salbach, J., Rachner, T., Rauner, M., Hempel, U., Anderegg, U., Franz, S., Simon, J.-C., Hofbauer, L., 2012. Regenerative potential of glycosaminoglycans for skin and bone. J. Mol. Med. 90, 625–635.

Sampath Kumar, N.S., Nazeer, R.A., Jaiganesh, R., 2012. Wound healing properties of collagen from the bone of two marine fishes. Int. J. Peptide Res. Therapeut. 18, 185–192.

Schnabelrauch, M., Scharnweber, D., Schiller, J., 2013. Sulfated glycosaminoglycans as promising artificial extracellular matrix components to improve the regeneration of tissues. Curr. Med. Chem. 20, 2501–2523.

Segura, T., Anderson, B.C., Chung, P.H., Webber, R.E., Shull, K.R., Shea, L.D., 2005. Crosslinked hyaluronic acid hydrogels: a strategy to functionalize and pattern. Biomaterials 26, 359–371.

Senni, K., Borchiellini, C., Duchesnay, A., Pellat, B., Letourneur, D., Kern, P., 1998. Antiproliferative polysaccharides modulate distribution and phenotypic expression of collagens by gingival fibroblasts. J. Biomed. Mater. Res. 40, 164–169.

Senni, K., Pellat, B., Gogly, B., Blondin, C., Letourneur, D., Jozefonvicz, J., Sinquin, C., Colliec-Jouault, S., Durand, P., 1999. Use of fucan for regulating the reconstruction of gingival and dermal connective tissues. US Patent 6559131. May 2003.

Senni, K., Gueniche, F., Yousfi, M., Fioretti, F., Godeau, G., Colliec-Jouault, S., Ratiskol, J., Sinquin, C., Raguenes, G., Courtois, A., Guezennec, J., 2005. Sulfated depolymerized derivatives of exopolysaccharides (EPS) from mesophilic marine bacteria, method for preparing same, and use thereof in tissue regeneration. US Patent 0131472. June 2008.

Senni, K., Gueniche, F., Foucault-Bertaud, A., Igondjo-Tchen, S., Fioretti, F., Colliec-Jouault, S., Durand, P., Guezennec, J., Godeau, G., Letourneur, D., 2006a. Fucoidan a sulfated polysaccharide from brown algae is a potent modulator of connective tissue proteolysis. Arch. Biochem. Biophys. 445, 56–64.

Senni, K., Sinquin, C., Colliec-Jouault, S., Godeau, G., Guezennec, J., 2006b. Use of a polysaccharide which is excreted by the *Vibrio diabolicus* species for the engineering of non-mineralized connective-tissue. US Patent 0317860. December 2008.

Senni, K., Pereira, J., Gueniche, F., Delbarre-Ladrat, C., Sinquin, C., Ratiskol, J., Godeau, G., Fischer, A.-M., Helley, D., Colliec-Jouault, S., 2011. Marine polysaccharides: a source of bioactive molecules for cell therapy and tissue engineering. Mar. Drugs 9, 1664–1681.

Senni, K., Gueniche, F., Changotade, S., Septier, D., Sinquin, C., Ratiskol, J., Lutomski, D., Godeau, G., Guezennec, J., Colliec-Jouault, S., 2013. Unusual glycosaminoglycans from a deep sea hydrothermal bacterium improve fibrillar collagen structuring and fibroblast activities in engineered connective tissues. Mar. Drugs 11, 1351–1369.

Song, A., Rane, A.A., Christman, K.L., 2012. Antibacterial and cell-adhesive polypeptide and poly(ethylene glycol) hydrogel as a potential scaffold for wound healing. Acta Biomater. 8, 41–50.

Song, E., Yeon, K.S., Chun, T., Byun, H.-J., Lee, Y.M., 2006. Collagen scaffolds derived from a marine source and their biocompatibility. Biomaterials 27, 2951–2961.

Tae, G., Kim, Y.J., Choi, W.I., Kim, M., Stayton, P.S., Hoffman, A.S., 2007. Formation of a novel heparin-based hydrogel in the presence of heparin-binding biomolecules. Biomacromolecules 8, 1979–1986.

Tanihara, M., Suzuki, Y., Yamamoto, E., Noguchi, A., Mizushima, Y., 2001. Sustained release of basic fibroblast growth factor and angiogenesis in a novel covalently crosslinked gel of heparin and alginate. J. Biomed. Mater. Res. 56, 216–221.

Taylor, K.R., Gallo, R.L., 2006. Glycosaminoglycans and their proteoglycans: host-associated molecular patterns for initiation and modulation of inflammation. FASEB J. 20, 9–22.

Teeri, T.T., Brumer III, H., Daniel, G., Gatenholm, P., 2007. Biomimetic engineering of cellulose-based materials. Trends Biotechnol. 25, 299–306.

Tombs, M.E.H.S., 1998. An introduction to polysaccharide biotechnology. In: Gunpowder Square, vol. 1. Taylor & Francis Publishers, London, UK.

Tu, Y., Mithieux, S.M., Annabi, N., Boughton, E.A., Weiss, A.S., 2010. Synthetic elastin hydrogels that are coblended with heparin display substantial swelling, increased porosity, and improved cell penetration. J. Biomed. Mater. Res. A 95A, 1215–1222.

Valappil, S.P., Misra, S.K., Boccaccini, A.R., Keshavarz, T., Bucke, C., Roy, I., 2007. Large-scale production and efficient recovery of PHB with desirable material properties, from the newly characterised *Bacillus cereus* SPV. J. Biotechnol. 132, 251–258.

van der Smissen, A., Hintze, V., Scharnweber, D., Moeller, S., Schnabelrauch, M., Majok, A., Simon, J.C., Anderegg, U., 2011. Growth promoting substrates for human dermal fibroblasts provided by artificial extracellular matrices composed of collagen I and sulfated glycosaminoglycans. Biomaterials 32, 8938–8946.

Van Vlierberghe, S., Dubruel, P., Schacht, E., 2011. Biopolymer-based hydrogels as scaffolds for tissue engineering applications: a review. Biomacromolecules 12, 1387–1408.

Warren, S.M., Fong, K.D., Nacamuli, R.P., Song, H.M., Fang, T.D., Longaker, M.T., 2003. Biomaterials for skin and bone replacement and repair in plastic surgery. Oper. Tech. Plast Reconstr. Surg. 9, 10–15.

Watanabe, J., Ishihara, K., 2003. Change in cell adhesion property on cytocompatible interface using phospholipid polymer grafted with poly(D,L-lactic acid) segment for tissue engineering. Sci. Technol. Adv. Mater. 4, 539–544.

Weiss, P., Vinatier, C., Guicheux, J., Grimandi, G., Daculsi, G., 2004. Self setting hydrogel as an extracellular synthetic matrix for tissue engineering. In: Barbosa, M.A.M.F.J.C.R.L.B. (Ed.), Bioceramics, vol. 16, pp. 1107–1110.

Yusof, N.L.B.M., Wee, A., Lim, L.Y., Khor, E., 2003. Flexible chitin films as potential wound-dressing materials: wound model studies. J. Biomed. Mater. Res. A 66A, 224–232.

Zarai, Z., Balti, R., Mejdoub, H., Gargouri, Y., Sayari, A., 2012. Process for extracting gelatin from marine snail (*Hexaplex trunculus*): chemical composition and functional properties. Process Biochem. 47, 1779–1784.

Liver tissue engineering using functional marine biomaterials

6

T. Jiang, B. Singh, Y.-J. Choi, T. Akaike, C.-S. Cho

6.1 Introduction

Tissue engineering using biomaterial scaffolds has fascinated many scientists and medical doctors because it aims to reconstruct living tissues for replacement of lost or damaged tissues of living organisms (Langer and Vacanti, 1993) and the engineered tissues can reduce or eliminate the need for organ replacement (Patrick et al., 1998). To achieve this goal, extracellular matrix (ECM) should be necessary for the cells to adhere, proliferate, and differentiate as they are in their native tissue (Rosso et al., 2004). Typically, artificial ECMs have been used in tissue engineering because they bring the desired cell types in contact with an appropriate environment, provide mechanical support until the newly formed tissues are structurally stable, and present and store bio-signaling molecules (Putnam and Mooney, 1996). The artificial ECMs can be classified into natural polymer-based ECMs and synthetic polymer-based ones (Kim et al., 2011). The natural polymer-based artificial ECMs such as proteins, proteoglycans, glycoproteins, and glycosaminoglycans are found in natural tissues and act structurally and functionally in the tissues (Rosso et al., 2005). The naturally derived polymers have excellent biological properties, such as cell adhesion, mechanical properties, biodegradability, and biocompatibility, than synthetic polymers. However, they have some limitations such as reproducibility on a large scale, complexity of purification, and difficulty in controlling degradation time (Badylak et al., 2009).

The design of natural polymer-based artificial ECMs for liver tissue engineering is very important because the liver plays important roles in our body, such as metabolism, storage, and synthesis and release of carbohydrates, vitamins, lipids, proteins, and detoxification, inactivation of endogenous and exogenous substances and activation of precursor molecules (Zakim and Boyer, 1990). Also, hepatocytes as the parenchymal cells among liver cells are anchorage-dependent cells and the hepatocytes lose their liver-specific functions and viability without an optimal artificial ECM (Chan et al., 2004).

This chapter covers the recent developments on liver tissue engineering using marine biomaterials such as chitosan, alginate, and pullulan (PU) among naturally derived polymers, which are biocompatible, low immunogenic, and low toxic. Moreover, mechanism of specific interaction between galactose moiety of the galactose-carrying biomaterials and asialoglycoprotein receptors (ASGPR) on the hepatocytes and importance of 3D ECM are explained.

Functional Marine Biomaterials. http://dx.doi.org/10.1016/B978-1-78242-086-6.00006-6
Copyright © 2015 Elsevier Ltd. All rights reserved.

6.2 Mechanism of specific interaction between galactose moiety of the galactose-carrying biomaterials and ASGPR on the hepatocytes

Pricer and Ashwell (1971) first identified that circulating asialoglycoproteins (ASGPs) bind to ASGPRs on the hepatocytes and then degraded. The hepatocytes possess cell surface receptors that recognize and bind molecules with exposed galactose, *N*-acetyl galactosamine, or glucose moieties (Neufeld et al., 1979). After binding to the receptors, internalization of ASGPs occurred through coated pits and coated vesicles and the tracers began to appear in a complex arrangement of larger smooth-surface vesicles (Wall et al., 1980). The hepatic plasma membrane receptors localized in the sinusoidal face of hepatocytes mediate specific binding and uptake of partially deglycosylated glycoproteins via receptor-mediated endocytosis (Ashwell and Harford, 1982). Therefore, many researches have attempted to imitate the process of incorporation of old proteins by the hepatocytes to mimic the biological environment.

6.3 Importance of 3D culture system

Traditionally, hepatocytes are seeded on 2D monolayer cultures using ECM-treated surfaces because this system have proved invaluable in studying basic liver biology in both practical and cost-effective methods (Mitalea, 1998). However, they cannot fully replicate and maintain convincing hepatocyte function *in vitro* and *in vivo* (Berthiaume et al., 1996). The 2D culture of hepatocytes forces the adherent cells to change their cytoskeleton toward flattened shapes and limits cell–cell and cell–ECM interactions, which consequently leads to reduced polarization and loss of signaling pathways necessary for normal functions of hepatocytes (Godoy et al., 2013). Generally, hepatocytes from 2D scaffold *in vitro* remain viable for only a few days and rapidly deviate from their differentiated phenotype (Gomez-Lechon et al., 1998). As a consequence, there is a strong demand for 3D hepatocyte culture for supplying oxygen and nutrients, and maintaining their functionality better than 2D culture. Especially, formation of multicellular hepatocyte spheroids in the 3D culture is a very powerful approach for enhancing liver-specific functions (Glicklis et al., 2000a).

There are two kinds of methods for making 3D scaffold. One of them is conventional method that includes fiber bonding (Gillett et al., 2011), phase separation (Ma and Zhang, 1999), particulate leaching (Holy et al., 2000), gas foaming (Son et al., 2011), freeze-drying (Yang et al., 2012), and melt molding (Oh et al., 2011). Among the methods, freeze-drying method of natural polymers has been widely used for liver tissue engineering because porosities and average pore sizes of the 3D ECM can be easily controlled by the cooling rate of the hydrated ECM (Chung et al., 2002a). On the other hand, the conventional method has several disadvantages, such as inconsistent procedures with shape limitations, being manual-based processes, and the use of toxic organic solvents and porogens. Other method includes computational ECM design (Hollister, 2005) and solid-free form (Kang and Cho, 2012), which both can overcome the disadvantages of the conventional manual-based manufacture methods. Each of these has advantages and disadvantages as shown in Table 6.1.

Table 6.1 Characteristics of 3D culture system

Technology	Advantages	Disadvantages
ECM sandwiching	Mimics aspects of native ECM Simple to use	Limited 3D organization Degradable Variable consistency
ECM hydrogels	Mimics aspects of native ECM Good cell–cell interaction	Mass transfer barriers Cell retrieval issues Degradable Variable consistency
Alginate sponges	Good cell–cell interaction Uses animal-free biomolecules Relatively easy to make	Isolated spheroid cultures Co-cultures difficult Mass transfer barriers Degradable
Spheroid cultures	Good cell–cell interaction Relatively easy to create	Cannot create complex structures Suffer from necrotic centers Organized co-cultures difficult Limited by spheroid size
Self-assembly peptide fibers	Mimics specific ECM peptide fibers Reproducible	Poor mechanical properties Relatively expensive to produce Degradable
Electrospun fibers	Versatile range of scaffolds available Inexpensive High porosity	Poor mechanical properties Poor scaffold consistency
Solvent casting/particle leaching scaffolds	Range of scaffolds available Inexpensive High porosity	Poor porosity control Poor scaffold consistency Residual contaminants
Gas-foaming scaffolds	Range of scaffolds available Inexpensive High porosity	Poor porosity control Poor scaffold consistency
3D printed scaffolds	Good porosity control Good scaffold consistency	Expensive machinery Limitations in scaffold architecture Low porosity

Source: Modified with Godoy et al. (2013)

6.4 Marine biomaterials as an artificial ECM

6.4.1 Chitosan, galactosylated chitosan, and chitosan hybrid

Chitosan obtained by deacetylation of chitin as a crystalline polysaccharide found in crustaceans and insects consist of repeating β(1→4)-linked D-glucosamine and *N*-acetyl-D-glucosamine as shown in Figure 6.1 (Onishi and Machida, 1999). Chitosan is one of the good candidates as an artificial ECM because it is biocompatible, biodegradable, minimal foreign body response, low immunogenic, and low toxic. However, it has some limitations as a scaffold such as no specific binding site to hepatocytes and difficulty in controlling the degradation and swelling properties (She et al., 2008). In this section, only 3D scaffolds of chitosan, galactosylated chitosan (GC), and chitosan hybrid system for liver engineering are explained.

6.4.1.1 Chitosan

First, Li et al. (2003) used porous chitosan scaffolds for culture of rat hepatocytes after rehydration procedure of lyophilized chitosan in sodium hydroxide solution. The prepared chitosan had a porosity of 90% with pore sizes from 50 to 200 μm. The hepatocyte attachment on the chitosan scaffold was much higher than on monolayer film exhibiting round morphologies of the cells. Besides, the modified chitosan scaffolds with alginate and heparin showed an increase of albumin secretion than nonmodified scaffolds.

Feng et al. (2010) prepared electrospun chitosan nanofiber for hepatocyte culture. The fibers exhibited a uniform diameter distribution of 112 nm and they were nontoxic, but they did not check functions of hepatocytes attached on nanofibers.

6.4.1.2 Galactosylated chitosan

Park et al. (2003) conjugated galactose moieties with chitosan as shown in Figure 6.2 to increase galactose-specific recognition between galactose of the GC and ASGPR in the hepatocytes. The adhered hepatocytes at high concentration of GC exhibited many spheroid formations after 24 h in the presence of epidermal growth factor (EGF), but they did not check cellular behaviors at 3D scaffolds.

Feng et al. (2009) fabricated GC nanofibrous scaffold by electrospinning technique for rat hepatocyte culture. The GC nanofibrous scaffolds had an average diameter of 160 nm and displayed slow degradation with suitable mechanical properties as an ECM for hepatocytes. The hepatocytes cultured on GC nanofibrous scaffold formed

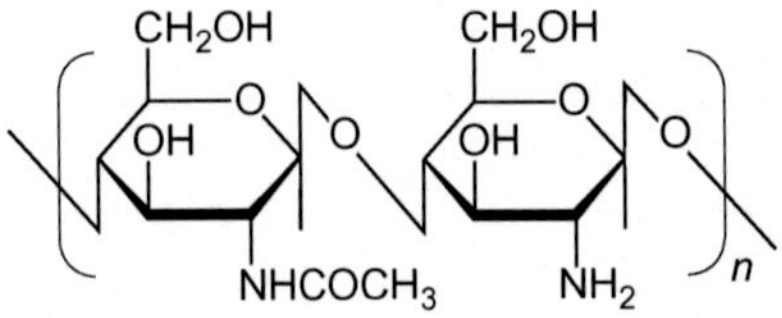

Figure 6.1 Chemical structures of chitosan.

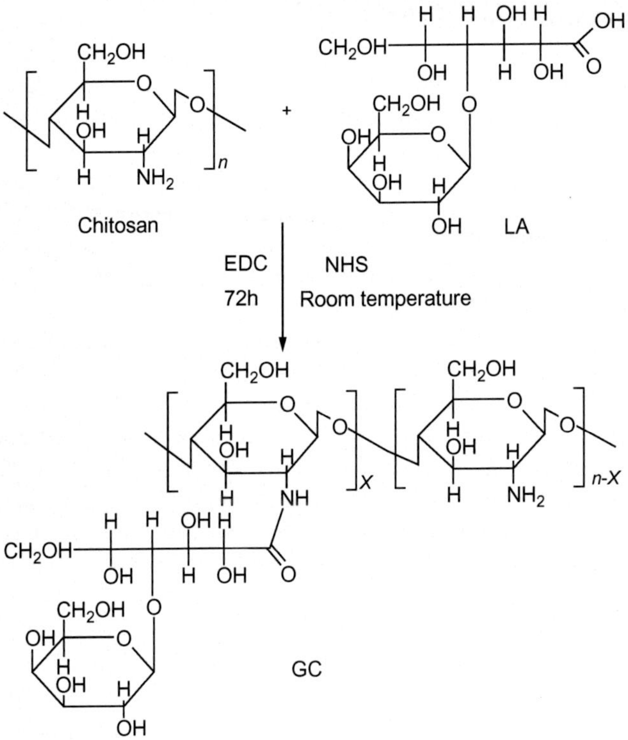

Figure 6.2 Synthesis scheme of GC.
Reprinted with permission from Park et al. (2003). Copyright © 2003, Elsevier.

spheroids of hepatocytes and exhibited higher albumin secretion, urea synthesis and cytochrome P-450 enzyme than on GC films, indicating the possibility of the scaffold for liver tissue engineering.

Kasoju and Bora (2012) fabricated GC/silk fibroin (SF) blend nanofibrous scaffold to mimic the biological/biochemical features of the natural ECM. Although they used HepG2 instead of primary hepatocytes, the growth, viability and metabolic activity of HepG2 on GC/SF scaffold were relatively higher than on GC or SF itself. Recently, Shang et al. (2014) prepared GC/hyaluronic acid (HA) blend sponge scaffolds for co-culture of hepatocytes and endothelial cells. HA is a ligand for endothelial cell-expressed CD44. The GC/HA scaffolds showed better biocompatibility, superior bioactivity, and higher liver-specific functions such as hepatocyte-specific gene expression, urea production, and drug metabolism than monoculture system.

6.4.1.3 Chitosan hybrid system

Risbud et al. (2003) coated textile scaffolds with chitosan/collagen blend for liver tissue engineering. The chitosan/collagen-coated scaffolds showed higher cell growth, more spheroid formation of hepatocytes and albumin secretion than chitosan-coated one. Similarly, Wang et al. (2005a) prepared ammonia-treated chitosan/collagen blend scaffolds to mimic the liver ECM. Hepatocytes cultured on the chitosan/collagen blend scaffold exhibited relatively high glutamate–oxaloacetate transaminase and glucose secretion functions for 25 days. They also added heparin into chitosan/collagen scaffold (Wang et al., 2005b). The chitosan/collagen/heparin scaffolds were highly porous, swelled rapidly in PBS solution and was stable *in vitro* for at least 60 days in collagenase/lysozyme solution. Moreover, hepatocytes cultured on the chitosan/collagen heparin scaffold showed highest urea and triglyceride secretion for 25 days.

Jiankang et al. (2009) prepared chitosan/gelatin blend scaffolds by a new fabrication process such as rapid prototyping, micro-replication, and freeze-drying to mimic well-organized specific architecture of natural liver. The scaffolds not only had analogous configurations of portal vein, central vein, flow-channel network, and hepatic chambers but also had high porosity (90%) with mean pore size of 100 μm. The hepatocytes cultured on the scaffolds formed large colonies in the predefined hepatic chambers and showed liver-specific functions such as albumin secretion and urea synthesis. Similarly, Gong et al. (2014) fabricated chitosan/gelatin hybrid scaffolds by indirect solid-free form and freeze-drying to create hierarchical channel network inside the scaffold. HepG2 cells attached on both micro-pores and micro-channels in the hybrid scaffolds. Further, HepG2 proliferated at much higher rate on the hybrid scaffolds with channel network than those without channels.

Kim and Rajagopalan (2010) prepared chitosan/HA multilayer scaffolds to mimic the space of Disse of liver and to preserve or enhance liver-specific behaviors of hepatocytes and liver sinusoidal endothelial cells (LSECs). Hepatocytes and LSECs cultured in the 3D hepatic model scaffold exhibited several key phenotypic characteristics for 12 days due to the enhanced heterotypic cell–cell interactions. Further, hepatocytes and LSECs on the chitosan/HA scaffolds increased CYP1A1/2 and CYP3A activity while albumin production increased threefold to sixfold.

She et al., (2008, 2009) prepared chitosan/SF hybrid scaffolds to increase biocompatibility owing to good biocompatibility of SF. The porosity of the scaffold was 95% when the total concentration of both was below 6 wt% with pore sizes of 100–150 μm. In addition, the scaffolds facilitated hepatocyte attachment and implanted scaffolds into omentum of rats showed only slight inflammation.

6.4.2 Alginate, galactosylated alginate, and alginate hybrid

Alginate as the structural component of brown seaweed is a linear polysaccharide of β (1 → 4)-linked L-guluronic acid and β (1 → 4)-linked D-mannuronic acid as shown in Figure 6.3 (Lee and Mooney, 2012). The alginate has been broadly used for biomaterials because it is biocompatible, nontoxic, easy to modify chemically, and especially undergoes gelation in the presence of divalent cations through ionic interaction. In this section, we discuss only 3D scaffolds of alginate, galactosylated alginate (GA), alginate hybrid, and alginate microcapsules (MCs) with other artificial ECMs.

6.4.2.1 Alginate

The most common method to prepare alginate MCs from an aqueous alginate solution is to mix the solution with divalent cations such as Ca^{++} and Ba^{++} as the cross-linking

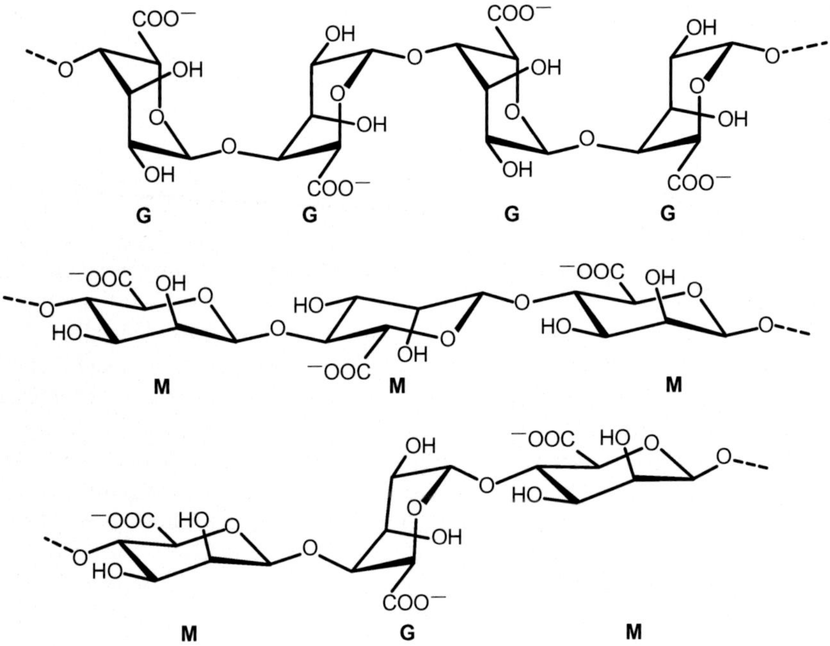

Figure 6.3 Chemical structures of G-block, M-block, and alternating block in alginate. Reprinted with permission from Lee and Mooney (2012). Copyright © 2012, Elsevier.

agents where only guluronate blocks of the alginate bind to divalent cations as the egg-box model of cross-linking (Grant et al., 1973).

Cohen and his coworkers investigated freezing regime during fabrication of the freeze-drying on the pore microstructure of alginate MCs (Zmora et al., 2002). The alginate MCs prepared by slowly freezing at −20 °C showed isotropic pore structure with spherical and interconnected pores, whereas the alginate ones obtained in liquid nitrogen with a temperature gradient showed two main regions of pore structure: small spherical pores at the interface with the cooling medium and above them a region with elongated pores. In the scaffolds with the isotropic spherical pores of the alginate MCs, the hepatocytes were clustered into spheroid formation. They also investigated the applicability of C3A as a human hepatocyte cell line instead of primary hepatocytes for predicting tool of drug metabolism after encapsulation of the cells into alginate MCs (Elkayam et al., 2006). The C3A cells cultured in alginate 3D scaffolds induced spheroid formation with enhanced drug metabolism activities such as cytochrome P-450 enzymes compared to that of 2D cultures. Furthermore, they cultured newborn rat liver cell isolates consisted of proliferating hepatocytes and progenitors into macroporous alginate MCs (pore size of 50–100 μm) to induce differentiation and maturation of the cells into functional hepatic tissue (Dvir-Ginzberg et al., 2008). The cells formed compacted spheroids within 3 days and secreted a high level of albumin, and the spheroids were developed into organoids after 6 weeks with an external mature hepatocyte layer covered by a laminin layer containing collagen, whereas the cells within collagen scaffolds did not express adult live enzymes or secrete albumin after 6 weeks.

Khattak et al. (2007) investigated effect of perfluorocarbons (PFCs) as oxygen carrier on metabolic activity and viability of HepG2 cells encapsulated into alginate MCs because the growth of cells after encapsulation is hindered by insufficient oxygen supply. The added PFCs into the alginate MCs increased cellular metabolic activity and cellular growth rate by 10%, and reduced both intracellular LDH and extracellular lactate levels by 20–40%.

Hwang et al. (2010) fabricated porous alginate MCs after loading of gelatin beads (size of beads: 150–300 μm) prepared by gelation of gelatin at 4 °C as a porogen into HepG2-encapsulated alginate MCs. Cell proliferation was enhanced under porous conditions and porous alginate MCs formed larger spheroid formation and higher albumin secretion than nonporous ones due to the enhanced mass transfer of nutrients, oxygen, and waste removal.

Gautier et al. (2011) investigated effect of alginate type and diameter of MCs on mass transfer and metabolic activities of encapsulated C3A cells into alginate MCs. The alginate MCs with a reduction of $CaCl_2$ concentration or gelation time led higher MCs porosity and better permeability. Moreover, the alginate MCs with a diameter of 1000 μm offered a reliable entrapment process for the cells; however, mechanical stability of the MCs should be considered and there were no differences of cell viability on metabolic activities by alginate type between low viscosity and medium one, and diameter of MCs between 600 and 1000 μm.

Bierwolf et al. (2012) seeded primary human hepatocytes isolated from the liver of metabolic-disordered children to recreate differentiated liver tissue-like spheroids on alginate MCs. Formation of hepatocyte spheroids was detected from day 3, and excellent metabolic functions of albumin, α1-antitrypsin, and urea were obtained with their maximum at day 7. Furthermore, highly differentiated hepatocytes in spheroids with a tissue-like structure on the alginate MCs was obtained, indicating that alginate MCs provide a favorable microenvironment for liver neo-tissue recreation and regeneration.

6.4.2.2 Galactosylated alginate

Although the alginate has several advantages such as biocompatibility, stability, adjustable porosity, and easy encapsulation of the cells, it has no specific ligand for the ASGPR of the hepatocytes. Therefore, chemical modifications might be considered to improve the specific interaction between the hepatocyte and the alginate.

Yang et al. (2002) coupled galactose moieties with alginate via the reaction of carboxylic acids of alginate and amine-modified lactobionic acid as shown in Figure 6.4 and evaluated liver functions of hepatocytes encapsulated into GA MCs. The liver function of hepatocytes in GA was higher than that of alginate ones due to the occurrence of binding of GA MCs with ASGPR of hepatocytes and spheroid formation of hepatocytes in the GA MCs.

Cho and his coworkers prepared alginate/GC scaffold by reaction of alginate gel with GC through the electrostatic interaction between alginate and GC to introduce galactose moieties into alginate scaffolds (Chung et al., 2002b). Porosity and pore size

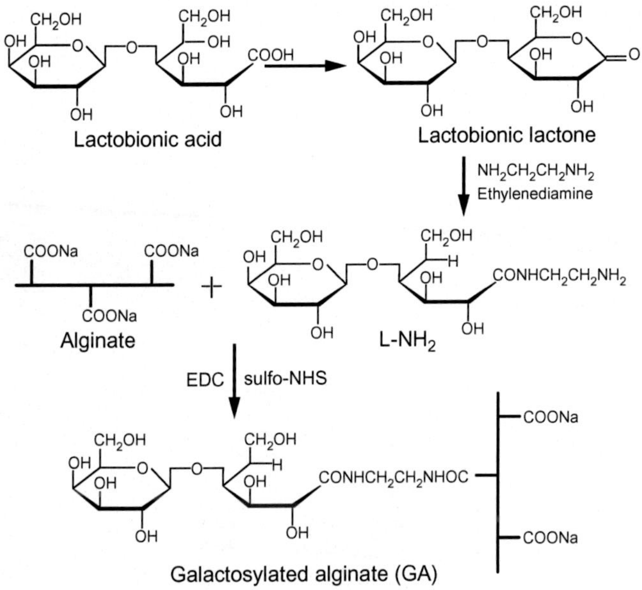

Figure 6.4 Reaction scheme of galactosylated alginate.
Reprinted with permission from Yang et al. (2002). Copyright © 2002, Elsevier.

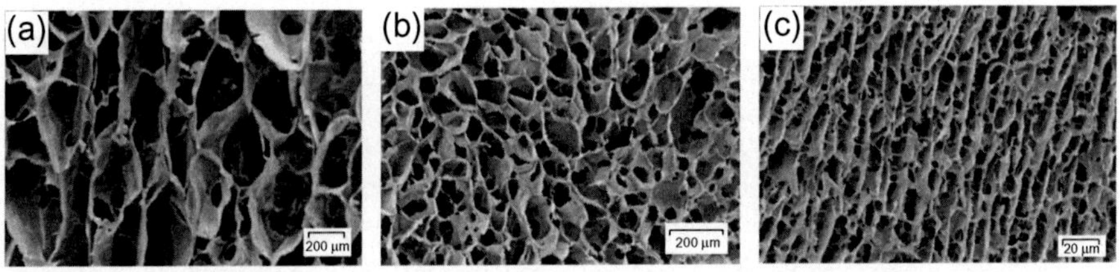

Figure 6.5 Scanning electron micrographs of a cross section of alginate/GC scaffolds as a function of the freezing temperature: (a) −20 °C, (b) −70 °C, and (c) liquid nitrogen.
Reprinted with permission from Chung et al. (2002b). Copyright © 2002, Elsevier.

of the alginate/GC scaffolds greatly depended on the molecular weight and content of GC and freezing temperature as shown in Figure 6.5. Moreover, the mechanical property of the alginate/GC scaffold increased with an increase of the GC content. Furthermore, spheroid formation of the hepatocytes for the alginate/GC scaffold was higher than those of alginate ones. They also co-cultured hepatocytes with NIH 3T3 in the alginate/GC scaffold to enhance liver-specific functions of the hepatocytes (Seo et al., 2006a). The co-culture of hepatocyte in alginate and alginate/GC scaffolds with NIH 3T3 resulted in enhanced liver-specific functions such as albumin secretion, ammonia elimination, and cytochrome P450(A) activity compared to those in hepatocyte monoculture as shown in Figure 6.6. Furthermore, they introduced heparin into the alginate/GC scaffold to enhance liver-specific functions (Seo et al., 2006b). The albumin secretion in the alginate/GC/heparin scaffold was markedly increased compared

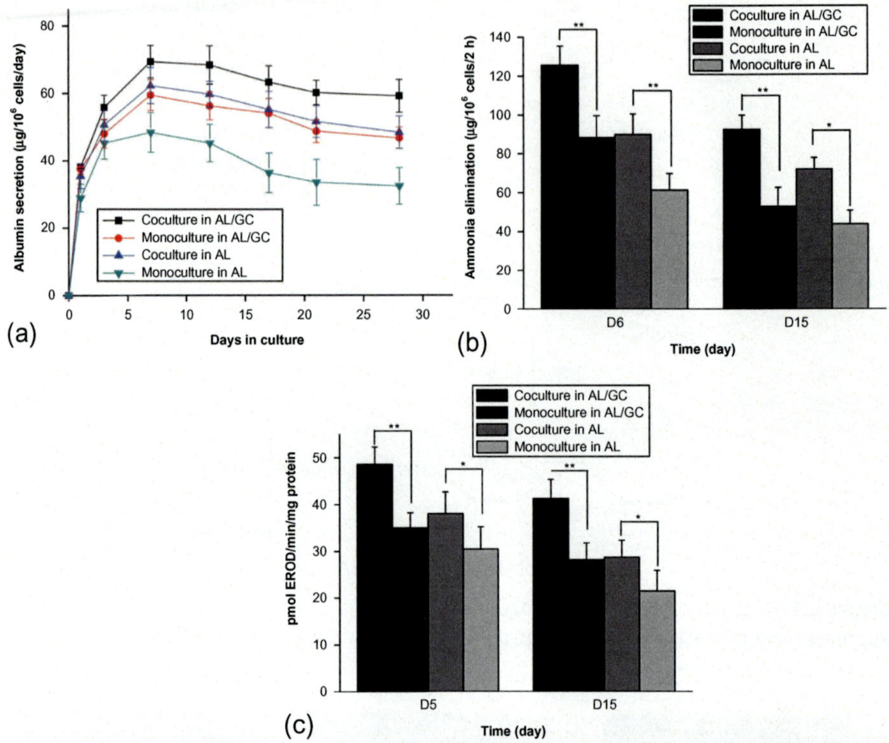

Figure 6.6 Comparison of liver-specific functions such as albumin secretion (a), ammonia elimination (b), and EROD activities (c) in AL and AL/GC sponges for monoculture and co-culture of hepatocytes with NIH3T3. (a) Albumin secretion rates for both systems were measured over 28 days in culture. ELISA specific for mouse albumin were used to quantify protein in 24 h conditioned medium. (b) Ammonia elimination was compared between both systems on day 5 and 16 of culture. (c) EROD deethylase activity by cytochrome P4501A1 was compared between both systems on day 6 and 16 of culture. EROD activity was measured using the highly fluorescent resorufin in both systems. All data are represented as mean $\pm$ $SD_{(n-1)}$ (n=6). Statistics (two-tailed paired Student's t-test) compared monoculture with co-culture; $^{*}P<0.05$ and $^{**}P_{o}<0.01$.
Reprinted with permission from Seo et al. (2006a). Copyright © 2006, Elsevier.

with that in alginate/GC one due to the provision of more multicellular spheroid formation mediated by cell-to-cell adhesion.

6.4.2.3 *Alginate hybrid system*

Alginate MCs prepared by ionic cross-linking between alginate and divalent cations have limitations, such as long-term stability in physiological conditions owing to ionic exchange reactions with monovalent cations and promotion of hemostasis *in vivo* by the released calcium ions from the MCs (Lee and Mooney, 2012). Therefore, cationic polymers such as poly (L-lysine) (PLL) and chitosan have been used to coat the alginate MCs that result in enhanced stability of the MCs.

Maguire et al. (2007) used PLL to coat alginate MCs and evaluated hepatic differentiation from the embryonic stem cells encapsulated into alginate/PLL MCs. It was found that a wide array of hepatocyte-specific markers was expressed by cells differentiated during a 23-day period within alginate/PLL MCs. In addition, spheroid formation was obtained in the MCs by E-cadherin protein-mediated differentiation.

6.4.2.4 Alginate MCs with other artificial ECMs

Hepatocytes encapsulated in alginate MCs can produce a very high cell density culture system with mechanical support to the cells as well as immunoprotection in the event of implantation. On the other hand, alginate matrix resists cell adhesion owing to the highly hydrated anionic surface of the alginate (Smetana, 1993), and cells do not attach to alginate matrix because cell–matrix interactions are mostly via receptors to the adsorbed proteins (Glicklis et al., 2000b). Therefore, other artificial ECMs for the hepatocytes can be entrapped into the alginate MCs to overcome the above-mentioned limitation.

Guo et al. (2003) added GC and poly (vinyl alcohol) (PVA) as artificial ECMs into hepatocyte-encapsulated alginate MCs. The hepatocytes encapsulated in the MCs showed more spheroid formation and higher metabolic function in albumin secretion than alginate MCs, suggesting that the added ECMs provide a better microenvironment for spheroid formation. Further, they added galactosylated poly (allyl amine) (GPA) and PVA as artificial ECMs into the hepatocyte-encapsulated alginate MCs (Kang et al., 2005). The hepatocytes cultured with GPA and PVA in the core of the alginate MCs were rapidly aggregated within a day and resulted in enhanced metabolic functions.

Seo et al. (2005) added xyloglucan (XG) as an artificial ECM into hepatocyte-encapsulated alginate MCs as the XG obtained from tamarind seed has galactose residues in the side chain. More than 70% of cultured hepatocytes in the alginate/XG scaffolds participated in spheroid formation after 2 days, whereas most hepatocytes in alginate MCs remained as a few aggregates even after 3 days as shown in Figure 6.7.

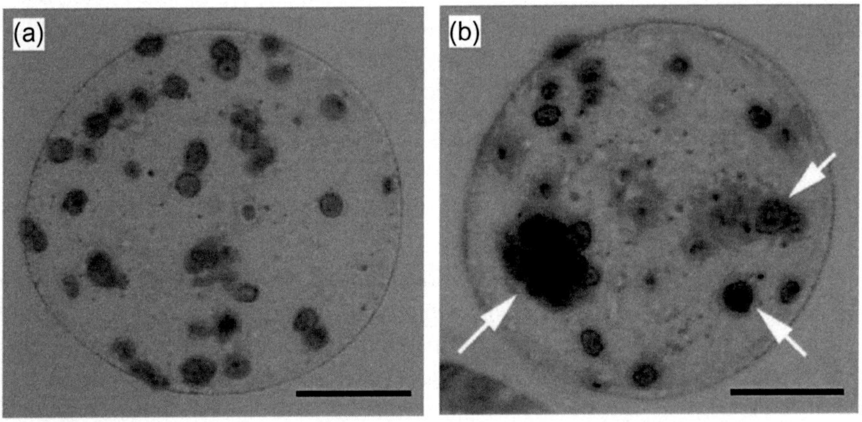

Figure 6.7 Morphology of hepatocytes cultured in AL (a) or AL/XG capsules (b). The hepatocytes in the capsules were cultured for 2 days within William's medium E supplemented with EGF and insulin; scale bar = 100 μm.

Reprinted with permission from Seo et al. (2005). Copyright © 2005, Elsevier.

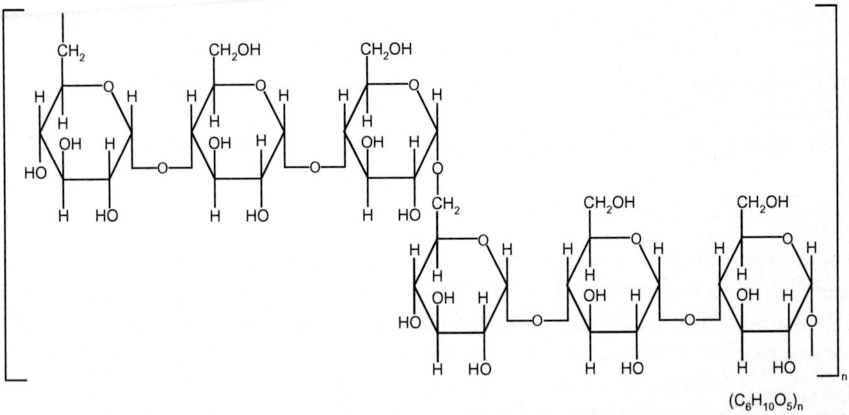

Figure 6.8 Structure of pullulan.

In addition, liver-specific functions of hepatocytes in the alginate/XG scaffolds were slowly decreased with culture time, whereas those were rapidly decreased in the alginate ones.

6.4.3 Pullulan

PU is produced from starch by *Aureobasidium pullulans* isolated from marine environments (Mulchandani et al., 1989). PU is a water-soluble, neutral linear polysaccharide consisting of α-1,6-linked maltotriose residues as shown in Figure 6.8 (Yuen, 1974). PU has been used for various biomedical applications such as gene delivery (Jo et al., 2007; Thakor et al., 2008) and wound healing (Prajapati et al., 2013), because it is biodegradable, hemocompatible, nonimmunogenic, and noncarcinogenic (Leathers, 2003). Among biomedical applications, PU has been investigated for use in various tissue engineering fields such as muscle (Autissier et al., 2007; Thebaud et al., 2007), bone (Hagashi et al., 2009; Fujioka-kobayashi et al., 2012), skin (Wong et al., 2011), and bladder (Abed et al., 2011). However, there has been no report of PU to use in liver tissue engineering, to our best knowledge. Since PU has high affinity to liver cells (Kaneo et al., 2001), the use of PU in liver tissue engineering is anticipated.

6.5 Summary

In this chapter, we explained summary of the works on liver tissue engineering using functional marine biomaterials including chitosan, alginate, and PU as artificial ECMs that provide cell-adhesive substrate, control 3D tissue structure and present growth factors, cell-adhesion signals and mechanical signals. In addition, galactose-carrying marine biomaterials regulate hepatocyte behaviors such as adhesion, morphology, proliferation, survival, migration, and differentiation through receptor-mediated interaction between ASGPR of hepatocytes and galactose moieties in the scaffolds. Especially, precise design of galactose-carrying marine biomaterials as artificial ECMs is required to develop bio-artificial liver-assist device or regeneration of the liver tissue substitutes for liver tissue engineering.

References

Abed, A., Assoul, N., Ba, M., Derkaoui, S.M., Portes, P., Louedec, L., Filaud, P., Bataille, I., Letourneur, D., Meddahi-Pelle, A., 2011. Influence of polysaccharide composition on the biocompatibility of pullulan/dextran-based hydrogels. J. Biomed. Mater. Res. Part A 96A, 535–542.

Ashwell, G., Harford, J., 1982. Carbohydrate-specific receptors of the liver. Annu. Rev. Biochem. 51, 531–544.

Autissier, A., Letourneur, D., Visage, C.L., 2007. Pullulan-based hydrogel for smooth muscle cell culture. J. Biomed. Mater. Res. 82 (A), 336–342.

Badylak, S.F., Fireytes, D.O., Gilbert, T.W., 2009. Extracellular matrix as a biological scaffold material: structure and function. Acta Biomater. 5, 1–13.

Berthiaume, F., Moghe, P.V., Toner, M., Yarmush, M.L., 1996. Effect of extracellular matrix topology on cell structure, function and physiological responsive ness: hepatocytes cultured in a sandwich configuration. FASEB J. 10, 1471–1484.

Bierwolf, J., Lutgehetmann, M., Deichmann, S., Erbes, J., Volz, T., Dandri, M., Cohen, S., Nashan, B., Pollok, J.M., 2012. Primary human hepatocytes from metabolic-disordered children recreate highly differentiated liver-tissue-like spheroids on alginate scaffolds. Tissue Eng. Part A 8, 1443–1453.

Chan, C., Berthiaume, F., Nath, B.D., Tilles, A.W., Toner, M., Yarmush, M.L., 2004. Hepatic tissue engineering for adjunct and temporary liver support critical technologies. Liver Transpl. 10, 1331–1342.

Chung, T.W., Yang, J., Akaike, T., Cho, K.Y., Nah, J.W., Kim, S.I., Cho, C.S., 2002a. Preparation of alginate/galactosylated chitosan scaffold for hepatocyte attachment. Biomaterials 23, 2827–2834.

Dvir-Ginzberg, M., Elkayam, T., Cohen, S., 2008. Induced differentiation and maturation of newborn live cells into functional hepatic tissue in macroporous alginate scaffolds. FASEB J. 22, 1440–1449.

Elkayam, T., Amitay-Shaprut, S., Dvir-Ginzberg, M., Harel, T., Cohen, S., 2006. Enhancing the drug metabolism activities of C3A-a human hepatocyte cell line by tissue engineering within alginate scaffolds. Tissue Eng. 12, 1357–1368.

Feng, Z., Chu, X., Huang, N., Wang, T., Wang, Y., Shi, X., Ding, Y., Gu, Z., 2009. The effect of nanofibrous galactosylated chitosan scaffolds on the formation of rat primary hepatocyte aggregates and the maintenance of liver function. Biomaterials 30, 2753–2763.

Feng, Z.Q., Leach, M.K., Chu, X.H., Wang, Y.C., Tian, T., Shi, X.L., Ding, Y.T., Gu, Z.Z., 2010. Electrospun chitosan nanofibers for hepatocyte culture. J. Biomed. Nanotechnol. 6, 658–666.

Fujioka-kobayashi, M., Ota, M.S., Shimoda, A., Nakahama, K., Akiyoshi, K., Miyamoto, Y., Iseki, S., 2012. Cholesteryl group and acryloyl group-bearing pullulan nanogel to deliver BMP2 and FGF18 for bone tissue engineering. Biomaterials 33, 7613–7620.

Gautier, A., Carpentier, B., Dufresne, M., Dinh, Q.V., Paullier, P., Legallais, C., 2011. Impact of alginate type and bead diameter on mass transfers and the metabolic activities of encapsulated C3A cells in bioartificial liver applications. Eur. Cell Mat. 21, 94–106.

Gillett, B.M., Rossen, N.S., Das, N., Leong, D., Wang, M., Dugar, A., Sia, S.K., 2011. Engineering extracellular matrix structure in 3D multiphase tissues. Biomaterials 32, 8067–8076.

Glicklis, R., Shapiro, L., Agbaria, R., Merchuk, J.C., Cohen, S., 2000a. Hepatocyte behavior within three-dimensional porous alginate scaffolds. Biotechnol. Bioeng. 67, 344–353.

Godoy, P., Hewitt, N.J., Albrecht, U., Andersen, M.E., Ansari, N., et al., 2013. Recent advances in 2D and 3D in vitro systems using primary hepatocytes, alternative hepatocyte sources and non-parenchymal liver cells and their use in investigating mechanisms of hepatotoxicity, cell signaling and ADME. Arch. Toxicol. 87, 1315–1530.

Gomez-Lechon, M.J., Jover, R., Donato, T., Ponsoda, X., Rodriguez, C., Stenzel, K.G., Klocke, R., Paul, D., Guillen, I., Bort, R., Castell, J.V., 1998. Long-term expression of differentiated functions in hepatocytes cultured in three-dimensional collagen matrix. J. Cell. Physiol. 177, 553–562.

Gong, H., Agustin, J., Wootton, D., Zhou, J.G., 2014. Biomimetic design and fabrication of porous chitosan-gelatin liver scaffolds with hierarchical channel network. J. Mater. Sci. Mater. Med. 25, 113–120.

Grant, G.T., Morris, E.R., Rees, D.A., et al., 1973. Biological interactions between polysaccharides and divalent cations: the egg-box model. FEBS Lett. 32, 195–198.

Guo, X.L., Yang, K.S., Hyun, J.Y., Kim, W.S., Lee, D.H., Min, K.E., Park, L.S., Seo, K.H., Kim, Y.I., Cho, C.S., Kang, I.K., 2003. Morphology and metabolism of Ba-alginate-encapsulated hepatocytes with galactosylated chitosan and poly (vinyl alcohol) as extracellular matrices. J. Biomater. Sci. Polym. Ed. 14, 551–565.

Hagashi, C., Hasegawa, U., Saita, Y., Hemmi, H., Hayata, T., Nakashima, K., Ezura, Y., Amagasa, T., Akiyoshi, K., Noda, M., 2009. Osteoblastic bone formation is induced by using nanogel-crosslinking hydrogel as novel scaffold for bone growth factor. J. Cell Physiol. 220, 1–7.

Hollister, S.J., 2005. Porous scaffold design for tissue engineering. Nat. Mater. 4, 518–524.

Holy, C.E., Shoichet, M.S., Davies, J.E., 2000. Engineering three dimensional bone tissue in vitro using biodegradable scaffolds. J. Biomed. Mater. Res. 51, 376–382.

Hwang, C.M., Sant, S., Masaeli, M., Kachouie, N.N., Zamanian, B., Lee, S.H., Khademhosseimi, A., 2010. Fabrication of three-dimensional porous cell-laden hydrogel for tissue engineering. Biofabrication 2, 035003, 12 pp.

Jiankang, H., Dichen, L., Yaxiong, L., Bo, Y., Hanxiang, Z., Qin, L., Bingheng, L., Yi, L., 2009. Preparation of chitosan-gelatin hybrid scaffolds with well-organized microstructures for hepatic tissue engineering. Acta Biomater. 5, 453.

Jo, J.I., Ikai, T., Dkazaki, A., et al., 2007. Expression profile of plasmid DNA by spermine derivatives of pullulan with different extents of spermine introduced. J. Control. Release 118, 389–398.

Kaneo, Y., Tanaka, T., Nakano, T., Yamaguchi, Y., 2001. Evidence for receptor-mediated hepatic uptake of pullulan in rats. J. Control. Release 70, 365–373.

Kang, H.W., Cho, D.W., 2012. Development of an indirect stereolithography technology for scaffold fabrication with a wide range of biomaterial selectivity. Tissue Eng. Part C Methods 18, 719–729.

Kang, I.K., Moon, J.S., Jeon, H.M., Meng, W., Kim, Y.I., Hwang, Y.J., Kim, S.K., 2005. Morphology and metabolism of Ba-alginate-encapsulated hepatocytes with galactosylated poly (allyl amine) and poly (vinyl alcohol) as extracellular matrices. J. Mater. Sci. Mater. Med. 16, 533–539.

Kasoju, N., Bora, U., 2012. Silk fibroin based biomimetic artificial extracellular matrix for hepatic tissue engineering applications. Biomed. Mater. 7, 045004 (1299).

Khattak, S.F., Chin, K.S., Bhatia, S.R., Roberto, S.C., 2007. Enhancing oxygen tension and cellular function in alginate cell encapsulation devices through the use of perfluorocarbons. Biotechnol. Bioeng. 96, 156–166.

Kim, Y., Rajagopalan, P., 2010. 3D hepatic cultures simultaneously maintain primary hepatocytes and liver sinusoidal endothelial cell phenotypes. PLoS ONE 5, e15456.

Kim, B.S., Park, I.K., Hoshiba, T., Jiang, H.L., Choi, Y.J., Akaike, T., Cho, C.S., 2011. Design of artificial extracellular matrices for tissue engineering. Prog. Polym. Sci. 36, 238–268.

Langer, R., Vacanti, J.P., 1993. Tissue engineering. Science 260, 920–926.

Leathers, T.D., 2003. Biotechnological production and applications of pullulan. Appl. Microbiol. Biotechnol. 62, 468–473.

Lee, K.Y., Mooney, D.V., 2012. Alginate: properties and biomedical applications. Prog. Polym. Sci. 37, 106–126.

Li, J., Pan, J., Zhang, L., Guo, X., Yu, Y., 2003. Culture of primary rat hepatocytes within porous chitosan scaffolds. J. Biomed. Mater. Rev. 67A, 938–943.

Ma, P.X., Zhang, R., 1999. Synthetic nano-scale fibrous extracellular matrix. J. Biomed. Mater. Res. 46, 60–72.

Maguire, T., Davidovich, A.E., Wallenstein, E.J., Novik, E., Sharma, N., Pedersen, H., Androulateis, I.P., Schloss, R., Yarmush, M., 2007. Control of hepatic differentiation via cellular aggregation in an alginate microenvironment. Biotechnol. Bioeng. 98, 631–644.

Mitalea, T., 1998. The current status of primary hepatocyte culture. Int. J. Exp. Pathol. 79, 393–409.

Mulchandani, A., Luong, J.H.T., Leduy, A., 1989. Biosynthesis of pullulan using immobilized *Aureobasidium pullulans* cells. Biotechnol. Bioeng. 33, 306–312.

Neufeld, E., Ashwell, G., Lennary, W. (Eds.), 1979. Carbohydrate recognizing systems for receptor-mediated pinocytosis. In: Biochemistry of Glycoprotein and Proteoglycans, vol. 33, Plenum Press, New York, pp. 241–266.

Oh, S.H., Park, S.C., Kim, H.K., Koh, Y.J., Lee, J.H., Lee, M.C., Lee, J.H., 2011. Degradation behavior of 3D porous polydioxanone-β-polycaprolactone scaffolds fabricated using the melt-molding particulate-leaching method. J. Biomater. Sci. Polym. Ed. 22, 225–237.

Onishi, H., Machida, Y., 1999. Biodegradation and distribution of water-soluble chitosan in mice. Biomaterials 20, 175–182.

Park, I.K., Yang, J., Jeong, H.J., Bom, H.S., Harada, I., Akaike, T., Kim, S.I., Cho, C.S., 2003. Galactosylated chitosan as a synthetic extracellular matrix for hepatocytes attachment. Biomaterials 24, 2331–2337.

Patrick, C.W., Mileos, A.G.M., McIntire, L.V. (Eds.), 1998. Frontiers in Tissue Engineering. Pergamon, Oxford.

Prajapati, V.D., Jani, G.K., Khanda, S.M., 2013. Pullulan: an exopolysaccharide and its various applications. Carbohydr. Polym. 95, 540–549.

Pricer, W.E., Ashwell, G., 1971. The binding of desialylated glycoproteins by plasma membranes of rat liver. J. Biol. Chem. 246, 4825–4833.

Putnam, A.J., Mooney, D.J., 1996. Tissue engineering using synthetic extracellular matrices. Nat. Med. 2, 824–826.

Risbud, M.V., Karamuk, E., Schlosser, V., Mayer, J., 2003. Hydrogel-coated textile scaffolds as candidate in liver tissue engineering: II. Evaluation of spheroid formation and viability of hepatocytes. J. Biomater. Sci. Polym. Ed. 14, 719–731.

Rosso, F., Giordano, A., Barbarisi, M., Barbarisi, A., 2004. From cell-ECM interactions to tissue engineering. J. Cell. Physiol. 199, 174–180.

Rosso, F., Marino, G., Giordano, A., Barbarish, M., Parmeggiani, D., Barbarisi, A., 2005. Smart materials as scaffolds for tissue engineering. J. Cell. Physiol. 203, 465–470.
Seo, S.J., Akaike, T., Choi, Y.J., Shirakawa, M., Kang, I.K., Cho, C.S., 2005. Alginate microcapsules prepared with xyloglucan as a synthetic extracellular matrix for hepatocyte attachment. Biomaterials 26, 3607–3615.
Seo, S.J., Kim, I.Y., Choi, Y.J., Akaike, T., Cho, C.S., 2006a. Enhanced liver functions of hepatocytes cocultured with NIH 3T3 in the alginate/galactosylated chitosan scaffold. Biomaterials 27, 1487–1495.
Seo, S.J., Choi, Y.J., Akaike, T., Higuchi, A., Cho, C.S., 2006b. Alginate/galactosylated chitosan/heparin scaffold as a new synthetic extracellular matrix for hepatocytes. Tissue Eng. 12, 33–44.
Shang, Y., Tamai, M., Ishii, R., Nagaoka, N., Yoshida, Y., Ogasawara, M., Yang, J., Tagawa, Y., 2014. Hybrid sponge comprised of galactosylated chitosan and hyaluronic acid mediates the co-culture of hepatocytes and endothelial cells. J. Biosci. Bioeng. 117, 99–106.
She, Z., Jin, C., Huang, Z., Zhang, B., Heng, Q., Xu, Y., 2008. Silk fibroin/chitosan scaffold: preparation, characterization, and culture with HepG2 cell. J. Mater. Sci. Mater. Med. 19, 3545–3553.
She, Z., Liu, W., Feng, Q., 2009. Self-assembly model, hepatocytes attachment and inflammatory response for silk fibroin/chitosan scaffolds. Biomed. Mater. 4, 045014.
Smetana, K., 1993. Cell biology of hydrogels. Biomaterials 14, 1046–1050.
Son, J.S., Kim, S.G., Oh, J.S., Appleford, M., Oh, S., Ong, J.L., Lee, K.B., 2011. Hydroxyapatite/polylactide biphasic combination scaffold loaded with dexamethasone for bone regeneration. J. Biomed. Mater. Res. A 99, 638–647.
Thakor, D.K., Teng, Y.D., Tabata, Y., 2008. Neuronal gene delivery by negatively charged pullulan-spermine/DNA anioplexes. Biomaterials 30, 1815–1826.
Thebaud, N.B., Pierron, D., Bareille, R., Visage, C.L., Letourneur, D., Bordenave, L., 2007. Human endothelial progenitor cell attachment to polysaccharide-based hydrogels: a pre-requisite for vascular tissue engineering. J. Mater. Sci. Mater. Med. 18, 339–345.
Wall, D.A., Wilson, G., Hubbard, A.L., 1980. The galactose-specific recognition system of mammalian liver: the route of ligand internaligation in rat hepatocytes. Cell 21, 79–93.
Wang, X., Yan, Y., Xiong, Z., Lin, F., Wu, R., Zhang, R., Lu, Q., 2005a. Preparation and evaluation of ammonia-treated collagen/chitosan matrices for liver tissue engineering. J. Biomet. Mater. Res. Part B Appl. Biomater. 75B, 91–98.
Wang, X., Yan, Y., Lin, F., Xiong, Z., Wu, R., Zhang, R., Lu, Q., 2005b. Preparation and characterization of a collagen/chitosan/heparin matrix for an implantable bioartificial liver. J. Biomater. Sci. Polym. Ed. 16, 1063–1080.
Wong, V.W., Rustad, K.C., Galvez, M.G., Neofytou, E., Glotzbach, J.P., Januszyk, M., Major, M.R., Sorkin, M., Longaker, M.T., Rajadas, J., Gurtner, G.C., 2011. Engineered pullulan-collagen composite dermal hydrogels improve early cutaneous wound healing. Tissue Eng. Part A 17, 631–644.
Yang, J., Goto, M., Ise, H., Cho, C.S., Akaike, T., 2002. Galactosylated alginate as a scaffold for hepatocytes entrapment. Biomaterials 23, 471–479.
Yang, X., Han, G., Pang, X., Fan, M., 2012. Chitosan/collagen scaffold containing bone morphogenetic protein-7 DNA supports dental pulp stem cell differentiation in vitro and in vivo. J. Biomed. Mater. Res. A.
Yuen, S., 1974. Pullulan and its applications. Process Biochem. 9, 7–9.
Zakim, D., Boyer, T.D. (Eds.), 1990. Hepatology: A Textbook of Liver Disease. W.B. Saunders Co, Philadelphia.
Zmora, S., Glicklis, R., Cohen, S., 2002. Tailoring the pore architecture in 3-D alginate scaffolds by controlling the freezing regime during fabrication. Biomaterials 23, 4081–4094.

Part Three

Applications of marine products in drug delivery systems and nano-pharmaceuticals

Functionalization of marine materials for drug delivery systems

7

S. Şenel

7.1 Introduction

A drug delivery system (DDS) can be defined as a formulation or device, which enables the introduction of a drug in the body and improves its efficacy and safety by controlling the rate, time, and site of its release. DDS is composed of two components. The first is the drug (active pharmaceutical ingredient, API), and the second is known as the excipient. API is defined as any component that is intended to furnish pharmacological activity or other direct effect in the diagnosis, cure, mitigation, treatment, or prevention of disease, or to affect the structure or any function of the body of man or other animals (EMA, 635046/2010). Excipients are the constituents of a pharmaceutical form that are not the active substance (EMA, 635046/2010). They can be classified as technological, applicatory, stabilizing, or biopharmaceutical excipients, according to their function.

The main intention of a DDS is to control the rate and site of absorption of the drug as well as its pattern of distribution and concentration within the body, and its duration of action. The routes of administration of these systems can be oral, parenteral, transmucosal, or transdermal. Yet, the current methods of delivery exhibit specific problems related to their limited or reduced absorption because of the partial degradation of the drug that occurs before they reach a desired target in the body. Therefore, targeted delivery systems have been designed to allow delivery of the drug intact to specifically targeted parts of the body without affecting other organs and tissues. During the past decade, polymeric micro- and nanoparticles, liposomes, polymer micelles, hydrogels, etc., which allow local delivery of therapeutics and diagnostics at disease sites, have been shown to be effective in enhancing drug targeting specificity, reducing systemic drug toxicity, improving absorption rates, and providing protection of drug against degradation.

Numerous natural and synthetic polymers have been investigated for development of such controlled release and targeted delivery systems. Natural polymers are good candidates for this purpose due to their biocompatibility and biodegradability (Silva et al., 2012). Polysaccharides have gained increasing attention as components of DDSs, particularly since they can be obtained in a well characterized and reproducible way from the natural sources. The field of natural polysaccharides of marine origin is large and expanding. Seaweeds are the most abundant source of polysaccharides such as alginates, agar and agarose, and carrageenans, which all have been widely investigated and utilized for drug delivery. The other polysaccharide of marine origin investigated for medical use is chitosan, which is extracted from the crab or shrimp

Functional Marine Biomaterials. http://dx.doi.org/10.1016/B978-1-78242-086-6.00007-8
Copyright © 2015 Elsevier Ltd. All rights reserved.

shells. New strategies including chemical modifications as well as combination of these polysaccharides with other polymers have been developed to improve or modify their properties.

In this chapter, after mentioning briefly about the properties of the two major marine materials (chitosan and alginate) utilized in delivery systems, DDSs, especially hydrogels and nanosystems, designed for controlled release or targeted delivery will be described and application of these marine polymers in such systems will be reviewed.

7.2 Chitosan

Chitosan is a cationic polysaccharide obtained by deacetylation of chitin, which is naturally available in crustaceans. It has a linear structure consisting of ß-(1-4)-linked D-glucosamine with randomly located *N*-acetylglucosamine groups depending upon the degree of deacetylation of the polymer (Figure 7.1). The degree of acetylation represents the proportion of *N*-acetyl-D-glucosamine units with respect to the total number of units and can be employed to differentiate between chitin and chitosan. Chitin with a degree of deacetylation of 65–70% or above is generally known as chitosan. The degree of deacetylation defines the physicochemical and biological properties of chitosan, which is specifically important in its applications. Chitosan has been favored over chitin due to its better solubility properties, and its free amine groups which are an active site in many chemical reactions. Chitosan is readily soluble in dilute acidic solutions below pH 6.0, and amino groups become deprotonated and the polymer loses its charge and becomes insoluble with the increasing pH. The solubility of chitosan is dependent on the degree of deacetylation and the method of deacetylation used since the pKa value is highly dependent on the degree of *N*-acetylation (Pillai et al., 2009).

The positive surface charge of chitosan allows it to interact with macromolecules like exogenous nucleic acids, negatively charged mucosal surfaces, or even the plasma membrane (Agnihotri et al., 2004; Kean et al., 2005). Chitosan is degraded by enzymes such as chitosanase and lysozyme. The rate of degradation of chitosan inversely depends on the degree of acetylation and crystallinity of the polymer. Distribution, degradation, and elimination processes of chitosan are reported to

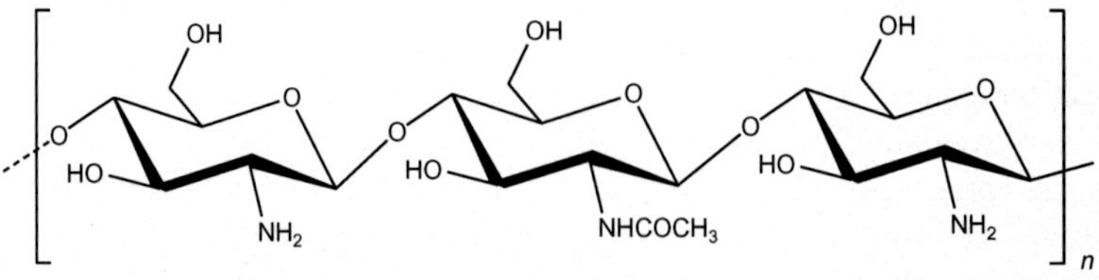

Figure 7.1 Chemical structure of chitosan.

be strongly dependent on its molecular weight and degree of deacetylation (Kean and Thanou, 2010).

In order to obtain desired physicochemical properties such as solubility, hydrophilicity, bioadhesivity, as well as improved bioactive properties, numerous derivatives of chitosan such as trimethylated, carboxymethylated, thiolated, and PEGylated chitosans, have been designed by chemical modifications.

It is possible to prepare different forms of formulation such as aqueous dispersions, suspensions, tablets, gels, films, sponges, patches, fibers, and particulate (nano and micro) systems using chitosan and its derivatives. Its versatility in physicochemical properties offers excellent opportunity to design custom-tailored delivery systems. Chitosan-based delivery systems for both conventional and modified release have been extensively investigated for oral, parenteral, mucosal, and transdermal drug/vaccine delivery due its properties such as biodegradability, biocompatibility, and bioadhesivity as well as its penetration-enhancing effect (Berger et al., 2004; Hejazi and Amiji, 2003; Sinha et al., 2004).

Chloride salt of chitosan is included in the European Pharmacopeia and recently, the base chitosan has been included in the United States Pharmacopeia.

7.3 Alginate

Alginate is extracted from some types of brown algae and is a linear anionic polysaccharide composed of 1-4-linked-β-D-mannuronic acid (M) and α-L-guluronic acid (G) residues (Figure 7.2). The ratio and the distribution of the M and G blocks significantly affect the affinity of alginate to pH and calcium ions because of the different relative positions of the carboxylic acid group in each block (Draget et al., 2003). The composition and extent of the sequences and the molecular weight determine the physical properties of the alginates. The presence of carboxylate groups within G units confers a global negative charge at pH 7. Gelation of alginate takes place by inducing the cross-linking of the alginate polymers with divalent cations (Tønnesen and Karlsen, 2002). The properties of the alginate gel are mainly influenced by the divalent cation content and the G-to-M ratio and sequence. Various methods of cross-linking have been introduced to precisely control the degradation process and mechanical properties (Holzapfel et al., 2013). The relatively mild gelation process has enabled proteins, cells, and DNA to be incorporated into alginate delivery systems with retention of full biological activity (Gombotz and Wee, 2012). Alginate exerts bioadhesive property, which makes it favorable especially for mucosal drug delivery.

Numerous alginate-based DDSs such as hydrogels, particulate systems, and sponges have been introduced for oral, mucosal, parenteral, and transdermal drug/vaccine delivery. Pharmaceutical grade of alginate is available according to its monographs in European Pharmacopeia and United States Pharmacopeia. Alginate also offers a wide range of possibilities of derivatization to improve its performance for biomedical applications.

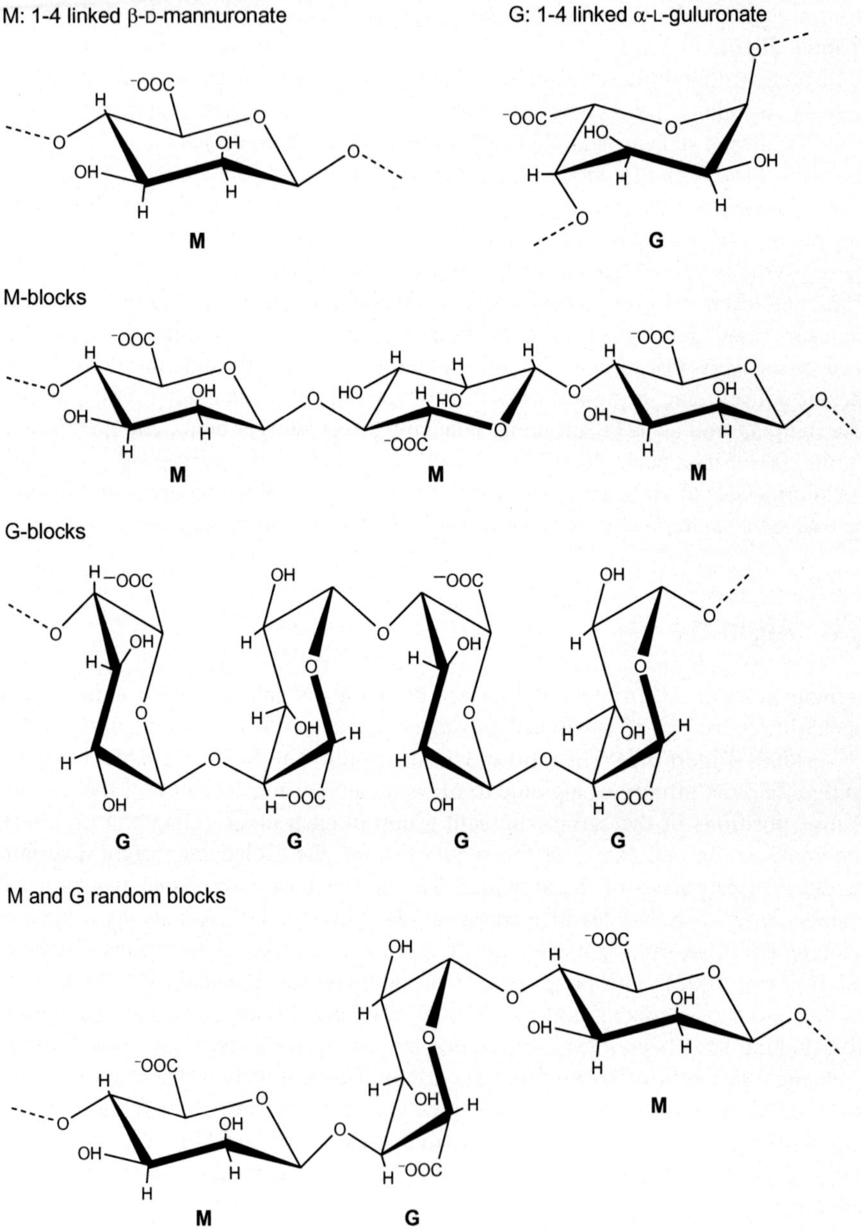

Figure 7.2 Monomers and block types of alginate.

7.4 Drug delivery systems

DDSs are developed to obtain a safe and effective drug concentration in body tissues. They are generally classified as conventional (immediate) release and modified release systems. Various terms including sustained release, prolonged release, extended release, delayed release, and controlled release are also employed to distinguish the modified release systems. They are particularly designed to control drug release rate, time, and site to further enhance the drug efficacy, safety, and compliance. It is also possible to direct the drug toward a specific organ or tissue, which is termed as targeted drug delivery. Targeted delivery systems release the drug at the desired site of action without any undesirable interaction at other sites and hence increase the therapeutic efficacy of the drug with minimum side effects. DDS must possess numerous features. The system should be recognized by the specific target tissues. The polymeric carrier itself, once in the targeting area should be able to control the drug administration by means of either a physiological or a chemical trigger. Polymers used must be biodegradable or readily eliminated from the body. The system should be stable both *in vitro* and *in vivo*. They should not induce modulation of disease state. Furthermore, the preparation of the DDS should be easy, reproducible, and cost-effective. In recent years, studies focused on stimuli-responsive DDSs to control and target the drug delivery (Cabane et al., 2012; Zhang et al., 2013). The drug release from these systems can be activated physically (e.g., magnetic field, electric field, ultrasound, temperature, and osmotic pressure), chemically (e.g., pH, ionic strength, and glucose), or biologically (enzymes and endogenous receptors) (Lopes et al., 2013).

Polymers play an essential role in the drug delivery technology by providing controlled release of drugs in constant doses over long periods, and tunable release of drugs. Colloidal systems such as micellar solutions, liquid crystal dispersions, as well as nanoparticle dispersions consisting of particles at nanosize have been shown to be very promising for polymeric DDSs with desired drug loading and release properties, long shelf-life, and low toxicity (Liechty et al., 2010).

Among the delivery systems, in which the marine polymers and their derivatives have been extensively utilized, are mainly hydrogels and nanosystems. Thus, in the following sections, these systems will be described in more detail.

7.4.1 Hydrogels

Hydrogels are hydrophilic polymer networks, which can absorb a significant amount of water. Hydrogels may be chemically stable or they may degrade and eventually disintegrate and dissolve (Hoffman, 2012; Li et al., 2012). The ability of these hydrogels in controlling water mobility makes them attractive for drug delivery (Mastropietro et al., 2012). The polymeric structure of hydrogels are held together by primary covalent cross-links, ionic forces, hydrogen bonds, affinity or "bio-recognition" interactions, hydrophobic interactions, polymer crystallites, physical entanglements of individual polymer chains, or a combination of two or more of the above interactions. They can be neutral hydrogels (uncharged), anionic (having negative charges only), cationic (having positive charges only), or ampholytic (having both positive and negative charges) (Peppas and Hoffman, 2013). Many synthetic and natural polymers can be used to form hydrogels for biomedical applications. Among them are also the marine polymers such as alginate and chitosan (George and Abraham, 2006).

It is also possible to use these polymers to form the interpenetrating networks in combination with polyacrylics such as poly(acrylic acid) and/or poly(acrylamide) resulting in multifunctional hybrid superporous hydrogels (d'Ayala et al., 2008; Laurienzo, 2010; Mastropietro et al., 2012). Their characteristics such as pH and temperature sensitivity can be modified to change the swelling capabilities and provide specific drug delivery. For example, a pH-sensitive hydrogel was produced by ionic cross-linking of chitosan with itaconic acid followed by polymerization and cross-linking with methacrylic acid (Milosavljevic et al., 2011). Swelling of chitosan-based superporous hydrogels was shown to be dependent on the pH and was reduced at higher cross-linking ratios (Park et al., 2006).

In recent years, microgels and nanogels became focus of attraction in drug delivery (Chacko et al., 2012; Kabanov and Vinogradov, 2009). Microgels can be defined as physically or chemically cross-linked hydrogel particles, which are confined to smaller dimensions (Pelton and Hoare, 2011). When the microgel particles are submicron-sized, they are known to be nanogels. Polysaccharide-based microgels/nanogels have high water content, functionality, biocompatibility, adjustable size from submicrons to nanometers, large surface area for multivalent bioconjugation, and interior network for the incorporation of therapeutics (Oh et al., 2009). Micro/nanogels, which are composed of hydrophilic or amphiphilic polymer chains, can be designed to spontaneously incorporate biologically active molecules through formation of salt bonds, hydrogen bonds, or hydrophobic interactions (Kabanov and Vinogradov, 2009). Polyelectrolyte nanogels can readily incorporate oppositely charged small and large molecules, such as siRNA, DNA, and proteins through electrostatic interaction (Zubareva et al., 2013). Multiple chemical functionalities can be employed in the nanogels for targeted drug delivery and imaging (Kabanov and Vinogradov, 2009).

Intelligent (smart) hydrogels, which can change their swelling behavior and other properties in response to environmental stimuli such as temperature, pH, solvent composition, and electric fields, have found wide applications in drug delivery. Ionic marine polysaccharides may exert biological functions in addition to their stimuli-responsive characters (Alvarez-Lorenzo et al., 2013; Prabaharan and Mano, 2006). Composites and grafted polymers can reinforce and extend their responsiveness.

Chemical modifications can be applied to polysaccharides to modify their final properties for improved drug delivery and targeting as well as for new biological activities. These strategies also involve the combination of these polysaccharides with other polymers (d'Ayala et al., 2008).

Chitosan can be chemically modified by grafting hydrophobic groups from amino groups of the main chains, where the proportion of the amine groups depends on the acetylation degree of the polymer.

Aqueous carbodiimide chemistry is widely used to couple carboxylic groups on alginate with molecules containing primary or secondary amines (d'Ayala et al., 2008). Several examples are reported in which alginate is used in combination with polyethyleneglycol (PEG). Chemical modification of alginate with acrylic acids through radical grafting of acrylic monomers can be performed to modulate the time of erosion and the rate of release of drugs (Laurienzo et al., 2006).

The relatively mild gelation process has enabled proteins, DNA to be incorporated into alginate matrices with retention of full biological activity. The pore size, degradation rate, and ultimately release kinetics can be controlled by the type of alginate and also coating agent. Gels of different morphologies can be prepared including large block matrices, large beads (>1 mm in diameter), microbeads, and nanobeads (Gombotz and Wee, 2012). However, alginate hydrogels cross-linked with divalent cations are unstable under conditions in which monovalent cations can be exchanged for the divalent cross-linking cations, leading to varying mechanical strengths and degradation profiles. It is possible to control the mechanical properties and degradation by conjugation of different functional groups to the carboxylates of alginate. It has also been possible to obtain responsive hydrogels based on temperature and pH by conjugation of different groups to alginate with subsequent divalent cation cross-linking (Alvarez-Lorenzo et al., 2013; Baldwin and Kiick, 2010; Matricardi et al., 2008).

Hybrid polymer networks of chitosan with alginate using a cross-linker has been shown to enable the covalent binding of the two macromolecules (Baysal et al., 2013). Recently, a novel "core-membrane" microgel drug delivery device for spontaneously pulsed release without any external trigger, which consisted of a chitosan–alginate microgel surrounded by a polyelectrolyte membrane formed by sodium polystyrene sulfonate and poly(allylamine hydrochloride), was reported (Zhou et al., 2013).

Aerogels with open pore structure and high surface area can be obtained from wet gels by using a suitable drying technology, usually supercritical drying process (Gaudio et al., 2013). There has been an increasing interest for the use of polysaccharide-based nanoporous aerogels for drug delivery (García-González et al., 2011; Veronovski et al., 2013). The size and morphology of aerogels can be customized by means of shaping of the gel by molding, extrusion, or any other suitable physical technique.

7.4.2 Nanodelivery systems

Nanodelivery systems can be defined as colloidal particles below 1000 nm size, such as nanospheres, nanocapsules, micelles, liposomes, solid lipid nanoparticles, nanocrystals, etc. The drug can be dissolved, entrapped, encapsulated, or attached to a nanoparticle matrix. Polymeric nanoparticle is a collective term given for any type of polymer nanoparticle (Soppimath et al., 2001). Generally, two main strategies are employed for their preparation: the dispersion of preformed polymers and the polymerization of monomers. Depending on the process used for the preparation of nanoparticles, nanospheres (matrix-type), or nanocapsules (reservoir-type) can be obtained (Couvreur, 2013; Fattal et al., 2012; Rao and Geckeler, 2011).

The control of drug release from the nanoparticles and their biodistribution as well as their ability to deliver nondiffusible compounds intracellularly is dependent on size, shape, mechanical property, and surface chemistry of the particles. Utilization of biodegradable polymers and surface functionalization of the nanoparticles has allowed selective delivery of therapeutic or diagnostic agents to the disease tissues and cells resulting in increased therapeutic activity and reduced toxicity (Couvreur, 2013).

Marine polymers that have been used for preparation of nanoparticles can be classified by their native charges such as cationic (chitosan) and anionic (alginate). Because of the presence of various derivable groups on their molecular chains, these polymers can be easily modified both chemically and biochemically. The hydrophilic groups, such as hydroxyl, carboxyl, and amino groups, of these polymers affect the polymer charge and can go under reaction with functional molecules. Polymer-based nanoparticles can be formed mainly by chemical or physical cross-linking, ionic complexation, and self-assembling (Nitta and Numata, 2013). Self-assembled particles and hydrogels can be formed by introduction of hydrophobic segments, such as deoxycholic acid, cholesterol, carboxylic acids, and hydrophobic polymers, into hydrophilic polysaccharide backbones. Depending on the polysaccharide/hydrophobic segments, molar ratios, and the lengths of both polysaccharides and hydrophobic segments, nanoparticles with different properties can be designed.

The most commonly used marine polysaccharide for nanoparticle preparation is chitosan as well as its derivatives (Garcia-Fuentes and Alonso, 2012; Jayakumar et al., 2010; Şenel, 2011). Chitosan-based nanoparticles have been investigated for controlled release and targeted delivery of low- and high-molecular-weight drugs for treatment of various disease conditions such as infectious diseases, diabetes, cancer, etc. (Arca et al., 2009; Goodarzi et al., 2013; Jean et al., 2012; Pandey and Khuller, 2004; Sayın et al., 2008; Sonia and Sharma, 2012). The positive charge of chitosan and its ability to form particulate systems without requiring harsh conditions of heat and/or organic solvents for incorporation of the drug offers advantages over other polymers used for this purpose (Sayın et al., 2009). Chemical cross-linking of chitosan has been achieved by covalent bonds between polysaccharide main chains via glutaraldehyde. Cationic, anionic, ampholytic, or uncharged cross-linked nanoparticles can be prepared by cross-linking chitosan using natural di- and tri-functional carboxylic acids as cross-linkers at different ratios.

Chitosan nanoparticles are also promising nonviral nucleotide (DNA, siRNA) delivery vector (Buschmann et al., 2013; Mao et al., 2010). They provide efficient intracellular delivery of nucleic acids by increasing the stability and also the intracellular uptake of the nucleic acids. Chitosan–nucleic acid nanoparticles may be readily formed by coacervation between the positively charged amine groups on the chitosan and negatively charged phosphate groups on the nucleotide, constituting a particular type of polyelectrolyte complexes (PECs). Molecular weight and degree of deacetylation of chitosan as well as its charge ratio affects the particle formation. Particle formation was shown to be dependent on the ratio of amine groups on chitosan to phosphate groups on DNA (Bowman and Leong, 2006; Rudzinski and Aminabhavi, 2010).

Chitosan as the cationic polysaccharide has been used with the anionic polysaccharides, carrageenan, and alginate for ionic complexation. Chitosan-carrageenan nanoparticles cross-linked with tripolyphosphate were prepared by corporation of polyelectrolyte complexation and ionic gelation (Grenha et al., 2010).

Grafting of the hydrophobic groups into amino groups of chitosan is possible via acylation by acyl chloride or acid anhydride, coupling with deoxycholic acid to design a new carrier for DNA delivery (Kim et al., 2005; Lee et al., 1998).

Modification of chitosan with polyglutamic acid and cell-binding peptide sequences has enabled the formulators to improve both drug and gene delivery (Casettari et al., 2012).

Although numerous derivatives of chitosan has been developed and shown to improve drug delivery and targeting, they must be comprehensively evaluated *in vivo* to ensure that the biocompatibility profile is retained.

Alginate is another example of marine polysaccharides that have been utilized for micro- and nanoparticle preparation. Calcium chloride is favorably utilized as a counter-ion for cross-linking nanogel formation. Alginate-based systems have been developed for delivery of different molecules such as insulin, anti-tubercular and anti-fungal drugs, etc. (Ahmad and Khuller, 2008; Haque et al., 2014). It has also been shown to be promising for gene delivery (Machado et al., 2013).

Alginate nanoparticles can be used as oral insulin carrier or glucose binder in the treatment of diabetes as a function of its chemical composition. High molecular weight M-rich alginate nanoparticles are a suitable vehicle for future development into oral insulin carrier (Kadir et al., 2013).

Several groups have developed chitosan–alginate complex systems at different sizes for delivery of wide range of drugs such as insulin (Goycoolea et al., 2009; Onishi et al., 2010). Alginate and chitosan can be used together to form drug-loaded modified release systems. The ionotropic gelation method and a slight modification in various ways are used to prepare these systems (beads) of different characteristics. The bead characteristics like morphology, buoyancy, swelling nature, drug entrapment efficiency, adsorption, and release behavior are of importance (Bhattarai et al., 2011). Insulin-loaded nanoparticles were prepared by alginate ionotropic pregelation followed by chitosan–polyelectrolyte complexation (Sarmento et al., 2007). Alginate inclusion through ionic gelation was shown to improve the vector properties of chitosan-based nanoparticles, demonstrating superior transfection ability while maintaining biocompatibility and low toxicity (Douglas et al., 2006).

7.5 Conclusion

Marine polymers possess several favorable features in comparison to synthetic polymers used in drug/gene/vaccine delivery, such as biocompatibility and biodegradability. Their physicochemical properties allow them to easily form different types of delivery systems such as nanoparticles and hydrogels, via suitable preparation methods, which do not affect the integrity and stability of the drugs/vaccines incorporated into them. Alginate and chitosan are the two main marine polymers used for drug delivery. It is possible to use them and their derivatives alone or in combination to develop an appropriate DDS with desired properties such as particle size, viscosity, swelling, release properties, etc. From the regulatory point of view, quality, safety, and efficacy are the three main requirements, which a DDS has to meet. For chitosan- and alginate-based delivery systems, their safety and efficacy have been demonstrated for numerous drugs and vaccines. In regard to quality, the standardization of these polymers is very important. For alginates, there are several alginate-based DDSs approved in the market, whereas there has been no chitosan-based DDS approved up to date yet. A formulator has to be aware of the fact that chitosan exerts bioactive properties; hence, when it is intended to be utilized as an excipient for a DDS, it may not be inert, and that is not accepted by the regulatory authorities. Nevertheless, chitosan is still a very attractive marine polymer for drug/gene/vaccine delivery.

References

Agnihotri, S.A., Mallikarjuna, N.N., Aminabhavi, T.M., 2004. Recent advances on chitosan-based micro- and nanoparticles in drug delivery. J. Control. Release 100 (1), 5–28.

Ahmad, Z., Khuller, G.K., 2008. Alginate-based sustained release drug delivery systems for tuberculosis. Expert Opin. Drug Deliv. 5 (12), 1323–1334.

Alvarez-Lorenzo, C., Blanco-Fernandez, B., Puga, A.M., Concheiro, A., 2013. Crosslinked ionic polysaccharides for stimuli-sensitive drug delivery. Adv. Drug Deliv. Rev. 65 (9), 1148–1171.

Arca, H.C., Günbeyaz, M., Şenel, S., 2009. Chitosan-based systems for the delivery of vaccine antigens. Expert Rev. Vaccines 8 (7), 937–953.

Baldwin, A.D., Kiick, K.L., 2010. Polysaccharide-modified synthetic polymeric biomaterials. Biopolymers 94 (1), 128–140.

Baysal, K., Aroguz, A.Z., Adiguzel, Z., Baysal, B.M., 2013. Chitosan/alginate crosslinked hydrogels: preparation, characterization and application for cell growth purposes. Int. J. Biol. Macromol. 59, 342–348.

Berger, J., Reist, M., Mayer, J.M., Felt, O., Peppas, N.A., Gurny, R., 2004. Structure and interactions in covalently and ionically crosslinked chitosan hydrogels for biomedical applications. Eur. J. Pharm. Biopharm. 57 (1), 19–34.

Bhattarai, R.S., Shrestha, A., Dhandapani, N.V., 2011. Drug delivery using alginate and chitosan beads: an overview. Chron. Young Sci. 2 (4), 192–196.

Bowman, K., Leong, K.W., 2006. Chitosan nanoparticles for oral drug and gene delivery. Int. J. Nanomedicine 1 (2), 117–128.

Buschmann, M.D., Merzouki, A., Lavertu, M., Thibault, M., Jean, M., Darras, V., 2013. Chitosans for delivery of nucleic acids. Adv. Drug Deliv. Rev. 65 (9), 1234–1270.

Cabane, E., Zhang, X., Langowska, K., Palivan, C.G., Meier, W., 2012. Stimuli-responsive polymers and their applications in nanomedicine. Biointerphases 7 (1–4), 9.

Casettari, L., Vllasaliu, D., Lam, J.K., Soliman, M., Illum, L., 2012. Biomedical applications of amino acid-modified chitosans: a review. Biomaterials 33 (30), 7565–7583.

Chacko, R.T., Ventura, J., Zhuang, J., Thayumanavan, S., 2012. Polymer nanogels: a versatile nanoscopic drug delivery platform. Adv. Drug Deliv. Rev. 64 (9), 836–851.

Couvreur, P., 2013. Nanoparticles in drug delivery: past, present and future. Adv. Drug Deliv. Rev. 65 (1), 21–23.

d'Ayala, G.G., Malinconico, M., Laurienzo, P., 2008. Marine derived polysaccharides for biomedical applications: chemical modification approaches. Molecules 13 (9), 2069–2106.

Douglas, K.L., Piccirillo, C.A., Tabrizian, M., 2006. Effects of alginate inclusion on the vector properties of chitosan-based nanoparticles. J. Control. Release 115 (3), 354–361.

Draget, K.I., Smidsrød, O., Skjåk-Bræk, G., 2003. Alginates from Algae. In: Steinbüchel, A. (Ed.), Biopolymers. John Wiley and Sons, Inc., Weinheim, pp. 215–224.

EMA (635046/2010). European Medicines Agency Glossary (terms and abbreviations). 9 December 2010.

Fattal, E., Hillaireau, H., Mura, S., Nicolas, J., Tsapis, N., 2012. Targeted delivery using biodegradable polymeric nanoparticles. In: Siepmann, J., Siegel, R.A., Rathbone, M.J. (Eds.), Fundamentals and Applications of Controlled Release Drug Delivery. Springer, USA, pp. 255–288.

Garcia-Fuentes, M., Alonso, M.J., 2012. Chitosan-based drug nanocarriers: where do we stand? J. Control. Release 161 (2), 496–504.

García-González, C.A., Alnaief, M., Smirnova, I., 2011. Polysaccharide-based aerogels—promising biodegradable carriers for drug delivery systems. Carbohydr. Polym. 86 (4), 1425–1438.

Gaudio, P.D., Auriemma, G., Mencherini, T., Porta, G.D., Reverchon, E., Aquino, R.P., 2013. Design of alginate-based aerogel for nonsteroidal anti-inflammatory drugs controlled delivery systems using prilling and supercritical-assisted drying. J. Pharm. Sci. 102 (1), 185–194.

George, M., Abraham, T.E., 2006. Polyionic hydrocolloids for the intestinal delivery of protein drugs: alginate and chitosan—a review. J. Control. Release 114 (1), 1–14.

Gombotz, W.R., Wee, S.F., 2012. Protein release from alginate matrices. Adv. Drug Deliv. Rev. 64, 194–205.

Goodarzi, N., Varshochian, R., Kamalinia, G., Atyabi, F., Dinarvand, R., 2013. A review of polysaccharide cytotoxic drug conjugates for cancer therapy. Carbohydr. Polym. 92 (2), 1280–1293.

Goycoolea, F.M., Lollo, G., Remunan-Lopez, C., Quaglia, F., Alonso, M.J., 2009. Chitosan-Alginate blended nanoparticles as carriers for the transmucosal delivery of macromolecules. Biomacromolecules 10 (7), 1736–1743.

Grenha, A., Gomes, M.E., Rodrigues, M., Santo, V.E., Mano, J.F., Neves, N.M., Reis, R.L., 2010. Development of new chitosan/carrageenan nanoparticles for drug delivery applications. J. Biomed. Mater. Res. A 92 (4), 1265–1272.

Haque, S., Md, S., Sahni, J.K., Ali, J., Baboota, S., 2014. Development and evaluation of brain targeted intranasal alginate nanoparticles for treatment of depression. J. Psychiatr. Res. 48 (1), 1–12.

Hejazi, R., Amiji, M., 2003. Chitosan-based gastrointestinal delivery systems. J. Control. Release 89 (2), 151–165.

Hoffman, A.S., 2012. Hydrogels for biomedical applications. Adv. Drug Deliv. Rev. 64, 18–23.

Holzapfel, B.M., Reichert, J.C., Schantz, J.T., Gbureck, U., Rackwitz, L., Noth, U., Jakob, F., Rudert, M., Groll, J., Hutmacher, D.W., 2013. How smart do biomaterials need to be? A translational science and clinical point of view. Adv. Drug Deliv. Rev. 65 (4), 581–603.

Jayakumar, R., Menon, D., Manzoor, K., Nair, S.V., Tamura, H., 2010. Biomedical applications of chitin and chitosan based nanomaterials—a short review. Carbohydr. Polym. 82 (2), 227–232.

Jean, M., Alameh, M., De Jesus, D., Thibault, M., Lavertu, M., Darras, V., Nelea, M., Buschmann, M.D., Merzouki, A., 2012. Chitosan-based therapeutic nanoparticles for combination gene therapy and gene silencing of in vitro cell lines relevant to type 2 diabetes. Eur. J. Pharm. Sci. 45 (1–2), 138–149.

Kabanov, A.V., Vinogradov, S.V., 2009. Nanogels as pharmaceutical carriers: finite networks of infinite capabilities. Angew. Chem. Int. Ed. Engl. 48 (30), 5418–5429.

Kadir, A., Mokhtar, M.T., Wong, T.W., 2013. Nanoparticulate assembly of mannuronic acid- and guluronic acid-rich alginate: oral insulin carrier and glucose binder. J. Pharm. Sci. 102 (12), 4353–4363.

Kean, T., Thanou, M., 2010. Biodegradation, biodistribution and toxicity of chitosan. Adv. Drug Deliv. Rev. 62 (1), 3–11.

Kean, T., Roth, S., Thanou, M., 2005. Trimethylated chitosans as non-viral gene delivery vectors: cytotoxicity and transfection efficiency. J. Control. Release 103 (3), 643–653.

Kim, K., Kwon, S., Park, J.H., Chung, H., Jeong, S.Y., Kwon, I.C., 2005. Physicochemical characterizations of self-assembled nanoparticles of glycol chitosan–deoxycholic acid conjugates. Am. Chem. Soc. 6, 1154–1158.

Laurienzo, P., 2010. Marine polysaccharides in pharmaceutical applications: an overview. Mar. Drugs 8 (9), 2435–2465.

Laurienzo, P., Malinconico, M., Mattia, G., Russo, R., La Rotonda, M.I., Quaglia, F., Capitani, D., Mannina, L., 2006. Novel alginate-acrylic polymers as a platform for drug delivery. J. Biomed. Mater. Res. A 78 (3), 523–531.

Lee, K.Y., Kwon, I.C., Kim, Y.H., Jo, W.H., Jeong, S.Y., 1998. Preparation of chitosan self-aggregates as a gene delivery system. J. Control. Release 51 (2–3), 213–220.

Li, Y., Rodrigues, J., Tomas, H., 2012. Injectable and biodegradable hydrogels: gelation, biodegradation and biomedical applications. Chem. Soc. Rev. 41 (6), 2193–2221.

Liechty, W.B., Kryscio, D.R., Slaughter, B.V., Peppas, N.A., 2010. Polymers for drug delivery systems. Annu. Rev. Chem. Biomol. Eng. 1, 149–173.

Lopes, J.R., Santos, G., Barata, P., Oliveira, R., Lopes, C.M., 2013. Physical and chemical stimuli-responsive drug delivery systems: targeted delivery and main routes of administration. Curr. Pharm. Des. 19 (41), 7169–7184.

Machado, A.H., Lundberg, D., Ribeiro, A.J., Veiga, F.J., Miguel, M.G., Lindman, B., Olsson, U., 2013. Encapsulation of DNA in macroscopic and nanosized calcium alginate gel particles. Langmuir 29 (51), 15926–15935.

Mao, S., Sun, W., Kissel, T., 2010. Chitosan-based formulations for delivery of DNA and siRNA. Adv. Drug Deliv. Rev. 62 (1), 12–27.

Mastropietro, D.J., Omidian, H., Park, K., 2012. Drug delivery applications for superporous hydrogels-review. Expert Opin. Drug Deliv. 9 (1), 71–89.

Matricardi, P., Meo, C.D., Coviello, T., Alhaique, F., 2008. Recent advances and perspectives on coated alginate microspheres for modified drug delivery. Expert Opin. Drug Deliv. 5 (4), 417–425.

Milosavljevic, N.B., Milasinovic, N.Z., Popovic, I.G., Filipovic, J.M., Krusic, M.T.K., 2011. Preparation and characterization of pH-sensitive hydrogels based on chitosan, itaconic acid and methacrylic acid. Polym. Int. 60 (3), 443–452.

Nitta, S.K., Numata, K., 2013. Biopolymer-based nanoparticles for drug/gene delivery and tissue engineering. Int. J. Mol. Sci. 14 (1), 1629–1654.

Oh, J.K., Lee, D.I., Park, J.M., 2009. Biopolymer-based microgels/nanogels for drug delivery applications. Prog. Polym. Sci. 34 (12), 1261–1282.

Onishi, H., Koyama, K., Sakata, O., Machida, Y., 2010. Preparation of chitosan/alginate/calcium complex microparticles loaded with lactoferrin and their efficacy on carrageenan-induced edema in rats. Drug Dev. Ind. Pharm. 36 (8), 879–884.

Pandey, R., Khuller, G.K., 2004. Polymer based drug delivery systems for mycobacterial infections. Curr. Drug Deliv. 1 (3), 195–201.

Park, H., Park, K., Kim, D., 2006. Preparation and swelling behavior of chitosan-based superporous hydrogels for gastric retention application. J. Biomed. Mater. Res. A 76A (1), 144–150.

Pelton, R., Hoare, T., 2011. Microgels and their synthesis: an introduction. In: Fernandez-Nieves, A., Wyss, H.M., Mattsson, J., Weitz, D.A. (Eds.), Microgel Suspensions: Fundamentals and Applications. Wiley-VCH, Singapore, pp. 3–32.

Peppas, N.A., Hoffman, A.S., 2013. Hydrogels. In: Ratner, B.D., Hoffman, A.S., Schoen, F.J., Lemons, J.E. (Eds.), Biomaterials Science, 3rd ed. Academic Press, Canada, pp. 166–179.

Pillai, C.K.S., Paul, W., Sharma, C.P., 2009. Chitin and chitosan polymers: chemistry, solubility and fiber formation. Prog. Polym. Sci. 34 (7), 641–678.

Prabaharan, M., Mano, J.F., 2006. Stimuli-responsive hydrogels based on polysaccharides incorporated with thermo-responsive polymers as novel biomaterials. Macromol. Biosci. 6 (12), 991–1008.

Rao, J.P., Geckeler, K.E., 2011. Polymer nanoparticles: preparation techniques and size-control parameters. Prog. Polym. Sci. 36 (7), 887–913.

Rudzinski, W.E., Aminabhavi, T.M., 2010. Chitosan as a carrier for targeted delivery of small interfering RNA. Int. J. Pharm. 399 (1–2), 1–11.

Sarmento, B., Ribeiro, A.J., Veiga, F., Ferreira, D.C., Neufeld, R.J., 2007. Insulin-loaded nanoparticles are prepared by alginate ionotropic pre-gelation followed by chitosan polyelectrolyte complexation. J. Nanosci. Nanotechnol. 7 (8), 2833–2841.

Sayın, B., Somavarapu, S., Li, X.W., Thanou, M., Sesardic, D., Alpar, H.O., Şenel, S., 2008. Mono-*N*-carboxymethyl chitosan (MCC) and *N*-trimethyl chitosan (TMC) nanoparticles for non-invasive vaccine delivery. Int. J. Pharm. 363 (1–2), 139–148.

Sayın, B., Somavarapu, S., Li, X.W., Sesardic, D., Şenel, S., Alpar, O.H., 2009. TMC-MCC (*N*-trimethyl chitosan-mono-*N*-carboxymethyl chitosan) nanocomplexes for mucosal delivery of vaccines. Eur. J. Pharm. Sci. 38, 362–369.

Şenel, S., 2011. Chitosan-based particulate systems for non-invasive vaccine delivery. Adv. Polym. Sci. 243, 111–137.

Silva, T.H., Alves, A., Popa, E.G., Reys, L.L., Gomes, M.E., Sousa, R.A., Silva, S.S., Mano, J.F., Reis, R.L., 2012. Marine algae sulfated polysaccharides for tissue engineering and drug delivery approaches. Biomatter 2 (4), 278–289.

Sinha, V.R., Singla, A.K., Wadhawan, S., Kaushik, R., Kumria, R., Bansal, K., Dhawan, S., 2004. Chitosan microspheres as a potential carrier for drugs. Int. J. Pharm. 274 (1–2), 1–33.

Sonia, T.A., Sharma, C.P., 2012. An overview of natural polymers for oral insulin delivery. Drug Discov. Today 17 (13–14), 784–792.

Soppimath, K.S., Aminabhavi, T.M., Kulkarni, A.R., Rudzinski, W.E., 2001. Biodegradable polymeric nanoparticles as drug delivery devices. J. Control. Release 70, 1–20.

Tønnesen, H.H., Karlsen, J., 2002. Alginate in drug delivery systems. Drug Dev. Ind. Pharm. 28 (6), 621–630.

Veronovski, A., Knez, Ž., Novak, Z., 2013. Preparation of multi-membrane alginate aerogels used for drug delivery. J. Supercrit. Fluids 79, 209–215.

Zhang, Y., Chan, H.F., Leong, K.W., 2013. Advanced materials and processing for drug delivery: the past and the future. Adv. Drug Deliv. Rev. 65 (1), 104–120.

Zhou, G., Lu, Y., Zhang, H., Chen, Y., Yu, Y., Gao, J., Sun, D., Zhang, G., Zou, H., Zhong, Y., 2013. A novel pulsed drug-delivery system: polyelectrolyte layer-by-layer coating of chitosan-alginate microgels. Int. J. Nanomedicine 8, 877–887.

Zubareva, A., Ily'ina, A., Prokhorov, A., Kurek, D., Efremov, M., Varlamov, V., Şenel, S., Ignatyev, P., Svirshchevskaya, E., 2013. Characterization of protein and peptide binding to nanogels formed by differently charged chitosan derivatives. Molecules 18 (7), 7848–7864.

Drug delivery systems and cartilage tissue engineering scaffolding using marine-derived products

8

T.H. Silva, R.L. Reis

8.1 The emergence of marine biomaterials in the biomedical arena

Scientists and engineers are constantly looking for new materials or combination of materials to develop biomedical systems with improved performance. Both synthetic and natural materials have allowed the production of different systems, with both presenting advantages and drawbacks. Normally, synthetic materials claim the possibility to control the chemistry and the degradability and, by their turn, natural materials exhibit better biocompatibility. Among them, polymeric materials have arisen as relevant bricks for the development of new biomaterials, with a high degree of variability, being thus tuneable in accordance with the envisaged application.

Among the natural origin materials, the ones produced from marine organisms have more recently found their place in the biomedical arena. In fact, although some marine biopolymers, such as chitosan or alginate, have been explored for several years; their majority only recently were more deeply addressed, namely, with the increasing attention that marine environment has received and with the technological advancements that allow the further exploration of such environment (Silva et al., 2012a). Marine origin materials in general and marine biopolymers in particular have been used in the development of drug delivery devices and tissue engineering systems, namely, scaffolds (Silva et al., 2012b). In the majority of the studies, following the development of tissue engineering in the early days, biomaterials are taken as matrices for supporting the culture of cells, that is, only with a mechanical role (Bae et al., 2014). However, their clinical application is facing some problems, related with the inadequate control over the cell's microenvironment (Bae et al., 2014). New technologies have been tackling this challenge and the development of functional biomaterials has been proposed by several authors (Sakiyama-Elbert and Hubbell, 2001; Ge et al., 2012; Bae et al., 2014). In functional biomaterials, the role of mechanical scaffold for cell culture is enhanced with the incorporation of bioactive agent(s) that influence cell fate and tissue regeneration. In fact, delivery systems can be incorporated in the tissue engineering scaffolds, by addition to the matrix or with the scaffold acting itself as a delivery system. The delivery of the bioactive compound(s) can then occur by diffusion, degradation of the matrix or by cell triggering (Sakiyama-Elbert and Hubbell, 2001). On the other hand,

Functional Marine Biomaterials. http://dx.doi.org/10.1016/B978-1-78242-086-6.00008-X
Copyright © 2015 Elsevier Ltd. All rights reserved.

micro and nanotechnologies have also emerged, allowing not only the fabrication of micro and nano drug delivery systems to be incorporated into the scaffolds but also to modify the surface properties of such scaffolds, namely, by adding topographical features aimed to control cell behavior (Bae et al., 2014).

With this chapter, the authors aim to illustrate the use of marine origin materials on the development of functional materials aimed for the controlled delivery of drugs and other bioactive agents, as well as for cartilage tissue regeneration. It is not intended to be a systematic review of the scientific literature in this domain, but to trigger the curiosity of the readers for the subject and to give some cues to know more about it.

8.2 Applications for controlled release of bioactive compounds

8.2.1 Requirements

From the given definition of functional biomaterials, drug delivery systems are commonly a component of such biomaterials. In this regard, there are some very good reviews of drug delivery systems and applications using marine origin materials, which we recommend for deeper understanding (Tabata and Ikada, 1998; Tonnesen and Karlsen, 2002; Kumar et al., 2004; Malafaya et al., 2007; Laurienzo, 2010; Silva et al., 2012b).

This section aims to illustrate some of the most common systems normally associated with a tissue regeneration goal. Thus, such systems are incorporated into the scaffolds or constitute the scaffolds themselves or aim at the delivery of factors affecting cell fate: growth factors or differentiation factors.

As abovementioned, the delivery of a drug or bioactive compound from a polymeric system can occur following different mechanisms, namely, diffusion, polymer(s) degradation and external stimuli (pH, temperature, cell triggering). In all the cases, it is aimed that the concentration of the compound on the delivery site is limited within a range of therapeutic action, as illustrated in the scheme of Figure 8.1. In this way, the release cannot be so slow that the compound will not exhibit relevant action (subtherapeutic level), but the higher concentrations, within the toxic level, should also be avoided for the higher prevalence of side effects.

With the number of factors affecting the delivery of the bioactive compounds, namely, the nature of the polymers constituting the system, the nature of the drug, the structural and chemical features of the system, among others, the development of an appropriate drug delivery system needs to be addressed in a case-by-case scenario. Below, several examples of drug delivery systems based in marine biopolymers will be assessed.

8.2.2 Particles and capsules

Chitosan is probably the most studied marine biopolymer for application in the development of drug delivery devices. More recently, the advent of chitosan-based nanomaterials has opened new possibilities in drug delivery, with particles exhibiting superior

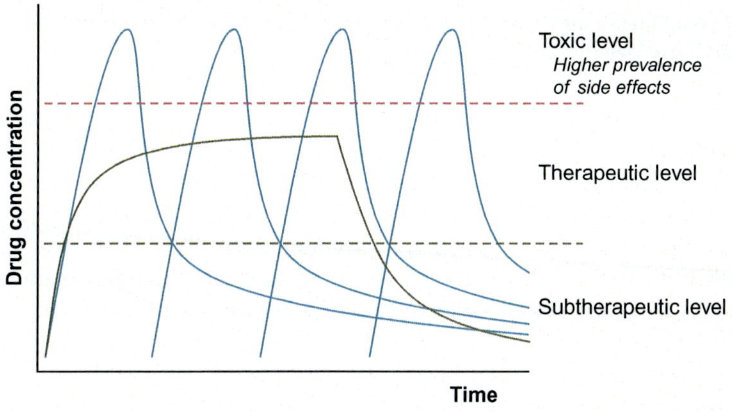

Figure 8.1 Illustrative scheme of the concentration of drugs along time in the periodic intake (blue) and when using a system for its sustained and controlled delivery (brown).

properties (Dash et al., 2011; Shukla et al., 2013). Chitosan nanoparticles have been prepared and incorporated with growth factors, not exhibiting *in vitro* cytotoxic or significant inflammatory behavior. Those growth factors, epidermal growth factor (EGF) and fibroblast growth factor (FGF) have been released in a sustained manner and in active form, demonstrated by the improvement in *in vitro* fibroblast proliferation as a consequence of adequate cellular signaling (Rajam et al., 2011). The delivery of bioactive compounds from chitosan nanoparticles can be also used as a tool for molecular imaging, allowing a noninvasive study of therapeutic proposals (Agrawal et al., 2010). Chitosan has been electrostatically complexed with chondroitin sulfate to form nanoparticles, able to deliver platelet lysates rich in growth factors, with relevant potential for bone tissue engineering (Santo et al., 2012). It has been also used as the derivative carboxymethylchitosan to modify the surface of synthetic dendrimers aimed to develop innovative systems for intracellular delivery of drugs (Oliveira et al., 2008). On the other hand, chitosan-based films and microparticles have been modified themselves with dendritic molecules, improving their properties as carriers for the controlled delivery of drugs (Aldana et al., 2010).

Algae biopolymers are also promising for the development of drug delivery devices. Alginate, obtained from brown algae, has been used on formulations of gastro-retentive dosage forms (Scholtz et al., 2014). Complex particulate systems based in alginate have been developed, aimed at insulin oral dosage. A blend of alginate, dextran sulfate, and chitosan processed as nanoparticles coated with albumin were able to retain insulin during up to 120 min in gastric simulated conditions in the absence of enzymes and then deliver as much as 80% of it in simulated intestinal fluid (Woitiski et al., 2009).

Carrageenans, sulfated polysaccharides isolated from red algae, has been also used for the development of polyelectrolyte complexes with chitosan, resulting in nanoparticles (Grenha et al., 2010) able to deliver drugs in a sustained way, with the potential to be uptaken by cells, thus exploring intracellular therapeutic approaches.

Hollow capsules can be also produced with marine origin biopolymers following a layer-by-layer strategy onto a sacrificial template, as is the case of capsules formed by hyaluronic acid with polyallylamine (Szarpak et al., 2008). Equivalent strategy has been also used to develop capsules with chitosan and elastin-like recombinamers (Costa et al., 2014). These capsules feature peptide sequences in their surface, bioactive RGD or nonfunctional RDG, proven to promote the viability of human mesenchymal stem cells. Additionally, they were also able to intracellularly deliver ovalbumin, promising application with other therapeutic compounds (Costa et al., 2014).

Table 8.1 illustrates some examples of particles and capsules based on marine biopolymers aimed at the controlled delivery of different types of drugs into several target tissues.

8.2.3 Membranes and hydrogels

Hydrogels based on chitosan can be produced by ion gelation, chemical crosslinking, or combination with other polymers (Silva et al., 2012b). Moreover, they can be also produced by chemical modification of chitosan, namely, by grafting peptides or polymers, as is the case of poly(ethylene glycol) sheet-shaped hydrogels (Ito et al., 2013). Illustrating the mixtures of polymers, a matrix of chitosan and alginate has been prepared with the incorporation of bone morphogenetic protein-2 and successfully used in the regeneration of bone tissue with enhanced results due to the release of the protein (Florezyk et al., 2013). Chitosan has been also combined with polycaprolactone on the development of nonwoven mats, with the presence of chitosan bearing antibacterial features to the mats. The incorporation of tea tree oil has an anti-inflammatory effect, which can be further enhanced by the inclusion of other active compounds, namely to explore other therapeutic effects (Bai et al., 2013). The combination of chitosan with bioactive compounds can, by its turn, result in gels with improved properties. For instance, the adding of hormone melatonin bearing antioxidant properties to a chitosan-based gel results in a functional biomaterial with anti-inflammatory activity to tackle oral mucositis (with high prevalence in patients receiving chemotherapy and irradiation in the head and neck) (Perchyonok et al., 2012a). The treatment of oral

Table 8.1 Examples of particulate systems from marine-based biopolymers in drug delivery applications

Polymers	Drug	Target	References
Alginate	–	–	Scholtz et al. (2014)
Alginate/chitosan	Insulin	Diabetes	Woitiski et al. (2009)
Chitosan/chondroitin sulfate	Platelet lysates	Bone regeneration	Santo et al. (2012)
Chitosan	EGF, FGF	Tissue regeneration	Rajam et al. (2011)
Chitosan	Dexamethasone	Bone regeneration	Oliveira et al. (2008)
Chitosan	Ovalbumin	–	Costa et al. (2014)
Chitosan/carrageenans	–	–	Grenha et al. (2010)
Hyaluronic acid	–	–	Szarpak et al. (2008)

mucositis has been also addressed by other chitosan-based membranes, in this turn combined with surfactants and propolis ethanolic extract with antioxidant capacity, capable of the sustained release of nystatin, an antifungal drug (Perchyonok et al., 2012b). Chitosan has been also combined with graphene oxide, which has been dispersed in the chitosan matrix by the establishment of amide bonds, resulting in films with higher mechanical properties and enhanced thermal stability (Zuo et al., 2013). Combination of chitosan with carbon nanotubes has been explored as well, although more focused in sensing and actuation (Lu and Chen, 2011). Chemical modification of chitosan is an alternating approach, explored for the improved processing of chitosan, as is the case of membranes prepared after the modification of chitosan with methyl acrylate and further crosslinked with glutaraldehyde, resulting in pH-sensitive films. Such membranes have been tested for the controlled delivery of ibuprofen, exhibiting higher resistance to the acidic stomach fluids, whereas the release is increased at pH 7.4 (Vieira et al., 2013). Other modifications, namely, attaching sugars, dendrimers, and cyclodextrins have been explored, with sugar-modified chitosan systems promising enhanced behavior on the controlled delivery of drugs (Sashiwa and Aiba, 2004). Additionally, chitosan can be also chemically modified for further combination with other polymers, strategy explored on the preparation of hydrogels with glycidyl methacrylate-chitosan and 2-(acryloyloxy) ethyl trimethylammonium, by photocrosslinking, which exhibit a pH and ionic strength-responsive behavior and are able to sustain the delivery of bovine serum albumin (BSA) (He and Chu, 2013).

Alginate has been successfully used in the development of functional biomaterials, namely, when coating collagen porous scaffolds to improve drug delivery applicability without significantly affecting the biological function of collagen scaffolds (Lee et al., 2012). Alginate hydrogels have been developed with preserved carboxylate moieties for stimuli-responsiveness and used for the protection of different proteins against the acidic gastric environment, allowing their later release at neutral pH (Chan and Neufeld, 2010). Nanofiber meshes have been prepared by electrospinning alginate, forming a hollow tube, in which a peptide-modified alginate hydrogel was injected for the sustained release of bone morphogenetic protein-2 and successfully used in guiding bone regeneration (Kolambkar et al., 2011). Nevertheless, alginate hydrogels face limited biodegradability and poor mechanical properties, which has motivated chemical scientists to synthesize new polymers based in poly(guluronate) to maintain the gelling favorable characteristic of alginate while addressing their limitations (Bouhadir et al., 1999). Other seaweed biopolymers are also used in the development of hydrogels for drug delivery, namely, the sulfated ones such as carrageenans. Kappa-carrageenan hydrogels have been supplemented with inorganic nanofillers, to control the release mechanism of methylene blue. This mechanism can be changed from diffusion to polymer relaxation due to the extent of swelling or variations in microstructure as a result of the incorporation of the nanofillers (da Silva et al., 2012).

Hyaluronic acid has been also obtained from marine organisms (Giji and Arumugam, 2014), namely, from the eyes of certain fish species, although nowadays the recombinant production is most probably the most explored technology to obtain hyaluronic acid for biomedical purposes. It has been used in the development of gels for drug delivery, namely, as dressings for the delivery of peptides (Rossi et al., 2012).

Table 8.2 Examples of hydrogel and membrane-like systems from marine-based biopolymers in drug delivery applications

Polymers	Drug	Target	References
Alginate	BMP-2	Bone regeneration	Kolambkar et al. (2011)
Alginate	Vitamin B12, lysozyme, insulin, albumin	–	Chan and Neufeld (2010)
Chitosan	–	–	Ito et al. (2013)
Chitosan	Ibuprofen	–	Vieira et al. (2013)
Chitosan	Tea tree oil	Inflammation	Bai et al. (2013)
Chitosan	–	–	Zuo et al. (2013)
Chitosan	Nystatin	Oral mucositis	Perchyonok et al. (2012b)
Collagen/alginate	–	Hard tissue regeneration	Lee et al. (2012)
Carrageenan	Methylene blue	–	da Silva et al. (2012)
Hyaluronic acid	Fluorescein isothiocyanate dextran	Vaginal mucosa	Rossi et al. (2012)
Hyaluronic acid	Pilocarpine	Ophthalmic	Vasi et al. (2014)

Table 8.2 condenses information on illustrative examples of membranes and gels based in marine biopolymers aimed at the controlled delivery of drugs.

8.3 Cartilage tissue engineering

8.3.1 Requirements

Articular cartilage is a connective tissue with a major role in physiological mobility, by allowing the frictionless movement of bones in a joint. It is composed mainly by hyaline cartilage, characterized by high content of collagen type II (distinguishing from fibrocartilage with collagen type I) and a matrix with several glycosaminoglycans, such as chondroitin sulfate. Hyaline cartilage can be distinguished from the third type (besides fibrocartilage) of cartilage, the elastic cartilage found in ear or nose, by the absence of elastin fibers. Although in small volume, chondrocytes are also present, being responsible by the expression of the extracellular matrix macromolecules. However, no vascular system or nerves are present, which turns cartilage hardly limited to self-heal. Thus, the loss or damage of cartilage as a consequence of trauma (accident, sport injury) or diseases, as osteoarthritis, is causing a dramatic disability in patients, with associated pain (Temenoff and Mikos, 2000; Ge et al., 2012, Matsiko et al., 2013).

Given the impact of such severe condition, scientists are pursuing different strategies to successfully repair or regenerate the lost tissue, but until now the ideal therapy is still being chased. This therapy would not only restore the frictionless movement on the joint but also support the repopulation with chondrocytes that will further produce an extracellular matrix completely integrated in the host tissue, thus resulting in a long-lasting effect (Mollon et al., 2013). The only FDA-approved cell-based therapy is autologous chondrocyte implantation to repair focal cartilage defects in young patients, while other approaches face limitations to induce the complete healing of cartilage defects, as is the case of microfracture, subchondral drilling, arthroscopic lavage, or abrasion arthroplasty (Ge et al., 2012). The use of autografts, mosaicplasty, or allografts presents important drawbacks as well as donor site morbidity and limited integration (Matsiko et al., 2013).

Tissue engineering approaches are being intensively investigated, relying on the development of an artificial biodegradable matrix that will serve as three-dimensional support (scaffold) for cell culture toward *de novo* cartilage tissue. However, the hierarchical nature of cartilage is quite challenging and cells need to receive the correct cues to regeneration of functional cartilage (and not fibrous tissue). In fact, the process should be considered as composed by three main factors: cells, a temporary extracellular matrix (scaffold), and growth factors (Mollon et al., 2013). This would mean a strategy in the called functional biomaterials, that is, not only with the support function as temporary extracellular matrix but also providing the cues for proper cell development. In addition, the multidisciplinarity required for the development of such complex systems supplemented by the contribution of clinical perspective would allow the achievement of better performing biomaterials (Johnstone et al., 2013). Nevertheless, most approaches are still based in scaffolds (as hydrogels or porous structures) without a particular bioactive performance. In this chapter, particular attention will be given to the development of functional scaffolds based in marine biopolymers. For the readers interested in extended information on the more common approaches, we recommend the reading of some good review papers, namely, the ones by Oliveira and Reis (2011), Balakrishnan and Banerjee (2011), or Popa et al. (2014).

8.3.2 Hydrogels and porous structures

Chitosan, mainly from shrimps and crabs, and alginate, from brown algae, are most probably the more used marine biopolymers on the development of structures tested as scaffolds in cartilage tissue engineering approaches. Nevertheless, as abovementioned, the examples of functional scaffolds based in marine biopolymers are still scarce, from which this field is still in its infancy. The more common approaches use scaffolds only as a temporary extracellular matrix and the biochemical cues are given by the addition of specific factors to the cell culture media. For instance, hydrogels of alginate and agarose has been used as three-dimensional support to study the chondrogenic differentiation of adipose-derived stem cells in culture medium containing transforming growth factor beta I, successfully observing the overexpression of proteins and proteoglycans (Awad et al., 2004).

Functional alginate hydrogels have been produced by the incorporation of tobacco mosaic virus to take advantage of its well-defined genetic/chemical modularity, multivalency, and well-defined structural features. It was demonstrated that according to the peptidic features exhibited by the viral particles, the cell attachment and differentiation could be adjusted (Luckanagul et al., 2012). Alginate beads have been also produced, in combination with hyaluronic acid, by ion gelation with calcium ions, in which sternal chondrocytes have been cultured. The mechanical agitation stimulus promoted the cell proliferation and metabolic activity, and in all cases, the nonarticular cells have been induced to synthesize articular-like matrix after 21 days of culture (Gerard et al., 2005). Additionally, interpenetrating networks of alginate and polyacrylamide have been prepared and incorporated into three-dimensional woven fiber scaffolds based in polycaprolactone, thus matching the inhomogenous properties of cartilage, while having a chondroinductive environment provided by the imbedded hydrogel. These biomimetic composite scaffolds revealed higher mechanical properties than any of their components alone and similar to the ones exhibited by cartilage, thus promising application as scaffolds in tissue engineering approaches or even as acellular replacement in cartilage repair (Liao et al., 2013). Three-dimensional plotting has been also employed to develop alginate-based scaffolds: pastes of alginate and poly (vinyl alcohol) were developed and extruded through the shell/core nozzles, resulting in a three-dimensional hollow fiber scaffold with well-oriented fibers. The higher concentrations of alginate used allow the production of more regular and stable structures, which exhibited mechanical properties equivalent to the ones presented by similar scaffolds and were able to support the attachment and spreading of human bone marrow-derived mesenchymal stem cells (hBMSCs) (Luo et al., 2013). This technique represents a simple methodology for the construction of complex structures, under mild and sterile conditions, thus promising high application in tissue regeneration approaches, namely, on the development of scaffolds with incorporation of growth and differentiation factors, as well as different cells, during the fabrication procedure.

Chitosan-based gels have been produced by different methodologies and proposed for the culture of chondrocytes or the chondrogenic differentiation of human mesenchymal stem cells. For instance, chitosan has been mixed with silk fibroin to obtain scaffolds supporting chondrocyte adhesion and proliferation, with expression of glycosaminoglycans and collagen type II (Zang et al., 2011). Chitosan gels have been also prepared by chemical cross-linking of solutions of chitosan, beta glycerol phosphate, and glucosamine with hydroxyethyl cellulose, which gel within minutes, adhere to cartilage and bone *in situ*, allowing their potential use as an arthroscopically injectable material for cartilage repair. The developed systems were able to support *in vitro* accumulation of cartilage matrix by primary articular chondrocytes, as well as *in vivo* when injected subcutaneously in nude mice. Moreover, the chitosan-based gel was also retained after injection in rabbit osteochondral defects, up to 1 week (Hoemann et al., 2005).

Nanofibrous membranes have been produced by electrospinning a solution of chitosan and polyethylene oxide together with poly(hexamethylene biguanide) hydrochloride, aiming to benefit from the antimicrobial properties of this compound. These membranes were produced with fibers ranging from 60 to 240 nm, depending on the concentration of the bioactive compound, and were shown to inhibit the growth of *Escherichia coli* and *Staphylococcus aureus* bacteria (Dilamian et al., 2013). Additionally, chitosan can be also used as a vector for the delivery of growth factors or functional genes, resulting in scaffolds with higher performance (Liu et al., 2011).

Besides the incorporation of biochemical factors, the functionality of the biomaterials can be given by other cues, such as morphological, mechanical, electric, or magnetic. In this perspective, micropatterned chitosan biomaterials have been developed by soft lithography, originating materials with grooves of defined width and depth, capable of influencing the spatial organization, spreading, and orientation of cells (Wang and Ho, 2004).

Another marine biopolymer that can also be used in cartilage regeneration approaches is chondroitin sulfate, a glycosaminoglycan that is a major constituent of cartilage extracellular matrix. Most approaches use the material extracted from terrestrial mammals, but similar biopolymers can be extracted from marine resources, namely, from cartilaginous fishes like sharks and rays, and thus similar strategies can be used. Besides their use to develop hydrogels, chondroitin sulfate can be also elegantly used as bioadhesive, by chemical functionalization with methacrylate and aldehyde groups, supporting the chemical link between the scaffold and the tissue proteins for improved integration (Wang et al., 2007).

Besides polysaccharides, resembling mainly the role of glycosaminoglycans in the cartilage extracellular matrix, proteins can be also used, in which collagen assumes an exciting role since besides the gelling possibilities, it contains sites for cellular adhesion and can influence chondrocyte differentiation of stem cells (Mollon et al., 2013). From the marine environment, most of the collagen that has been extracted is collagen type I, namely, from fish skins (Silva et al., 2014), and this type of collagen is characteristic of fibrous cartilage and not of articular cartilage, in which type II collagen is the major component. Probably the most known source of type II collagen in the marine environment is jellyfish and Gelinsky and co-workers have recently described the development of jellyfish collagen scaffolds for cartilage regeneration (Hoyer et al., 2014). Pepsin-soluble collagen extracted from *Rhopilema esculentum* (jellyfish species) was refibrillized with Tris buffer in the presence of sodium chloride at neutral pH and then used to prepare scaffolds by freeze-drying, further submitted to chemical crosslinking. These marine collagen-based scaffolds did not reveal any toxic effects over human mesenchymal stem cells, which were able to undergo chondrogenic differentiation under appropriate stimulation, as demonstrated by the overexpression of chondrogenic markers, such as collagen type II, aggrecan, and sulfated glycosaminoglycans, which are also representative of the envisaged cartilage extracellular matrix.

Table 8.3 depicts information on examples of hydrogels and porous structures based in marine biopolymers aimed at application as scaffolds in cartilage tissue regeneration strategies.

8.4 Final remarks

Marine environment has been receiving growing attention over the last years in which concerns the discovery of new chemical entities or new raw materials for the production of known compounds, in efforts of the so-called marine (or blue) biotechnology. Among those efforts, the development of marine biomaterials is giving the first steps toward a supported and stable role in controlled drug delivery and tissue regeneration.

Table 8.3 Examples of hydrogels and porous structures from marine-based biopolymers as cartilage tissue engineering scaffolds

Polymers	Processing route	Cells	References
Agarose	Gelation	hASCs	Awad et al. (2004)
Alginate	Ion gelation	hASCs	Awad et al. (2004)
Alginate/hyaluronic acid	Ion gelation	Nonarticular chondrocytes	Gerard et al. (2005)
Alginate/polyacrylamide	Interpenetrating gel	–	Liao et al. (2013)
Alginate	Ion gelation	Stem cells	Luckanagul et al. (2012)
Alginate	3D plotting	hBMSCs	Luo et al. (2013)
Carrageenan	Ion gelation	hASCs, ATDC5	Popa et al. (2012)
Carrageenan	Chemical crosslinking	hASCs	Pirraco et al. (2013)
Collagen	Freeze-drying	hMSCs	Hoyer et al. (2014)
Chitosan	Soft lithography	Endothelial cells	Wang and Ho (2004)
Chitosan	Particle agglomeration	hASCs	Malafaya et al. (2005)
Chitosan	Chemical crosslinking	Articular chondrocytes	Hoemann et al. (2005)
Chitosan/PEO	Electrospinning	–	Dilamian et al. (2013)
Chitosan/silk fibroin	Freeze-drying	Rabbit ear chondrocytes	Zang et al. (2011)

It benefits from the advancements made with equivalent terrestrial natural products, which definitely reduced some bottlenecks, but need to face the difficulties of a new area trying to find a place among established ones. In this sense, a significant role of marine biomaterials will be dependent on significant achievements: more appropriate materials, because are easier to produce, cheaper, safer, or other; better performing biomaterials, with improved *in vitro* results and successful *in vivo* experiments in adequate animal models to act as landmarks of the potential of marine biomaterials. These will set the basis to elaborate the dossier for regulatory approval to proceed to clinical trials, from which therapeutic solutions could finally arise. We will have to wait probably 10–20 years to see it. Working to shorten this period, several institutions are now looking to marine biomaterials through a multidisciplinary perspective and synergic attitude, designing methodologies to produce innovative biomaterials learning the lessons from the past and the ones constantly (and often surprisingly) given by Nature.

In the particular case of drug delivery, the advent of nanotechnology seems to represent a critical aspect, with several nanoparticulate systems being currently proposed based in marine biopolymers. This will be most probably the future, although there is still a big challenge to face: safety. Regarding the case of cartilage regeneration, the ideal solution seems to rely on a bioactive scaffold combined with cells. The cellular part, not addressed in this chapter, is a challenge, precisely due to the almost acellular nature of articular cartilage. Even though considering that issue adequately tackled by using chondrogenic differentiation of stem cells, there are still the challenges of the development of effective scaffolds, meaning structures that are capable to support the cell attachment and proliferation to produce the adequate extracellular matrix components, induce proper differentiation, without promoting significant immune response when implanted. Although the examples are still reduced when compared with the approaches in general and marine biomaterials can be considered as an area still in its childhood, chitosan and alginate are well-established materials and one believe that those two, together with the recently "hot" marine biopolymers collagens and sulfated polysaccharides, will play an important and leading role on the therapeutic solutions based in natural materials. Let's work together to achieve it!

Acknowledgments

The authors would like to acknowledge the funding from ERDF through POCTEP Project 0687_NOVOMAR_1_P and from European Union's Seventh Framework Programme (FP7/2007-2013) under grant agreement no. REGPOT-CT2012-316331-POLARIS.

References

Agrawal, P., Strijkers, G.J., Nicolay, K., 2010. Chitosan-based systems for molecular imaging. Adv. Drug Deliv. Rev. 62, 42–58.

Aldana, A.A., Martinelli, M., Strumia, M., 2010. Synthesis and properties of dendronized chitosan. Macromol. Symp. 298, 99–107.

Awad, H.A., Wickham, M.Q., Leddy, H.A., et al., 2004. Chondrogenic differentiation of adipose-derived adult stem cells in agarose, alginate, and gelatin scaffolds. Biomaterials 25, 3211–3222.

Bae, H., Chu, H., Edalat, F., et al., 2014. Development of functional biomaterials with micro- and nanoscale technologies for tissue engineering and drug delivery applications. J. Tissue Eng. Regen. Med. 8, 1–14.

Bai, M.Y., Chou, T.C., Tsai, J.C., Yang, H.C., 2013. Active ingredient-containing chitosan/polycaprolactone nonwoven mats: characterization and their functional assays. Mater. Sci. Eng. C Mater. Biol. Appl. 33, 224–233.

Balakrishnan, B., Banerjee, R., 2011. Biopolymer-based hydrogels for cartilage tissue engineering. Chem. Rev. 111, 4453–4474.

Bouhadir, K.H., Hausman, D.S., Mooney, D.J., 1999. Synthesis of cross-linked poly(aldehyde guluronate) hydrogels. Polymer 40, 3575–3584.

Chan, A.W., Neufeld, R.J., 2010. Tuneable semi-synthetic network alginate for absorptive encapsulation and controlled release of protein therapeutics. Biomaterials 31, 9040–9047.

Costa, R.R., Girotti, A., Santos, M., et al., 2014. Cellular uptake of multilayered capsules produced with natural and genetically engineered biomimetic macromolecules. Acta Biomater. 10, 2653–2662.

da Silva, A.L.D., Salgueiro, A.M., Fateixa, S., et al., 2012. Swelling and release properties of functional kappa-carrageenan hydrogel nanocomposites. MRS Proc. 1403, 213–219.

Dash, M., Chiellini, F., Ottenbrite, R.M., Chiellini, E., 2011. Chitosan – a versatile semi-synthetic polymer in biomedical applications. Prog. Polym. Sci. 36, 981–1014.

Dilamian, M., Montazer, M., Masoumi, J., 2013. Antimicrobial electrospun membranes of chitosan/poly(ethylene oxide) incorporating poly(hexamethylene biguanide) hydrochloride. Carbohydr. Polym. 94, 364–371.

Florezyk, S.J., Leung, M., Li, Z.S., et al., 2013. Evaluation of three-dimensional porous chitosan-alginate scaffolds in rat calvarial defects for bone regeneration applications. J. Biomed. Mater. Res. A 101, 2974–2983.

Ge, Z., Li, C., Heng, B.C., et al., 2012. Functional biomaterials for cartilage regeneration. J. Biomed. Mater. Res. A 100A, 2526–2536.

Gerard, C., Catuogno, C., Amargier-Huin, C., et al., 2005. The effect of alginate, hyaluronate and hyaluronate derivatives biomaterials on synthesis on non-articular chondrocyte extracellular matrix. J. Mater. Sci. Mater. Med. 16, 541–551.

Giji, S., Arumugam, M., 2014. Isolation and characterization of hyaluronic acid from marine organisms. Adv. Food Nutr. Res. 72, 61–77.

Grenha, A., Gomes, M.E., Rodrigues, M., et al., 2010. Development of new chitosan/carrageenan nanoparticles for drug delivery applications. J. Biomed. Mater. Res. A 92A, 1265–1272.

He, M.Y., Chu, C.C., 2013. Dual stimuli responsive glycidyl methacrylate chitosan-quaternary ammonium hybrid hydrogel and its bovine serum albumin release. J. Appl. Polym. Sci. 130, 3736–3745.

Hoemann, C.D., Sun, J., Légaré, A., et al., 2005. Tissue engineering of cartilage using an injectable and adhesive chitosan-based cell-delivery vehicle. Osteoarthritis Cartilage 13, 318–329.

Hoyer, B., Bernhardt, A., Lode, A., et al., 2014. Jellyfish collagen scaffolds for cartilage tissue engineering. Acta Biomater. 10, 883–892.

Ito, T., Yoshida, C., Murakami, Y., 2013. Design of novel sheet-shaped chitosan hydrogel for wound healing: A hybrid biomaterial consisting of both PEG-grafter chitosan and crosslinkable polymeric micelles acting as drug containers. Mater. Sci. Eng. C Mater. Biol. Appl. 33, 3697–3703.

Johnstone, B., Alini, M., Cucchiarini, M., et al., 2013. Tissue engineering for articular cartilage repair – the state of the art. Eur. Cell. Mater. 25, 248–267.

Kolambkar, Y.M., Dupont, K.M., Boerckel, J.D., et al., 2011. An alginate-based hybrid system for growth factor delivery in the functional repair of large bone defects. Biomaterials 32, 65–74.

Kumar, M.N.V.R., Muzzarelli, R.A.A., Muzzarelli, C., et al., 2004. Chitosan chemistry and pharmaceutical perspectives. Chem. Rev. 104, 6017–6084.

Laurienzo, P., 2010. Marine polysaccharides in pharmaceutical applications: an overview. Mar. Drugs 8, 2435–2465.

Lee, H.J., Ahn, S.H., Kim, G.H., 2012. Three-dimensional collagen/alginate hybrid scaffolds functionalized with a drug delivery system (DDS) for bone tissue regeneration. Chem. Mater. 24, 881–891.

Liao, I.-C., Moutos, F.T., Estes, B.T., Zhao, X., Guilak, F., 2013. Composite three-dimensional woven scaffolds with interpenetrating network hydrogels to create functional synthetic articular cartilage. Adv. Funct. Mater. 23, 5833–5839.

Liu, X., Ma, L., Mao, Z.W., et al., 2011. Chitosan-based biomaterials for tissue repair and regeneration. In: Jayakumar, R., Prabaharan, M., Muzzarelli, R.A.A. (Eds.), Chitosan for Biomaterials II (Book Series: Advanced in Polymer Science). Springer, Heidelberg, pp. 81–127.

Lu, L.H., Chen, W., 2011. Supramolecular self-assembly of biopolymers with carbon nanotubes for biomimetic and bio-inspired sensing and actuation. Nanoscale 3, 2412–2420.

Luckanagul, J., Lee, L.A., Nguyen, Q.L., et al., 2012. Porous alginate hydrogel functionalization with virus as three-dimensional scaffolds for bone differentiation. Biomacromolecules 13, 3949–3958.

Luo, Y., Lode, A., Gelinsky, M., 2013. Direct plotting of three-dimensional hollow fiber scaffolds based on concentrated alginate pastes for tissue engineering. Adv. Healthcare Mater. 2, 777–783.

Malafaya, P.B., Pedro, A.J., Peterbauer, A., et al., 2005. Chitosan particles agglomerated scaffolds for cartilage and osteochondral tissue engineering approaches with adipose tissue derived stem cells. J. Mater. Sci. Mater. Med. 16, 1077–1085.

Malafaya, P.B., Silva, G.A., Reis, R.L., 2007. Natural-origin polymers as carriers and scaffolds for biomolecules and cell delivery in tissue engineering applications. Adv. Drug Deliv. Rev. 59, 207–233.

Matsiko, A., Levingstone, T.J., O'Brien, F.J., 2013. Advanced strategies for articular cartilage defect repair. Materials 6, 637–668.

Mollon, B., Kandel, R., Chahal, J., et al., 2013. The clinical status of cartilage tissue regeneration in humans. Osteoarthr. Cartil. 21, 1824–1833.

Oliveira, J.M., Kotobuki, N., Marques, A.P., et al., 2008. Surface engineered carboxymethylchitosan/poly(amidoamine)dendrimer nanoparticles for intracellular targeting. Adv. Funct. Mater. 18, 1840–1853.

Oliveira, J.T., Reis, R.L., 2011. Polysaccharide-based materials for cartilage tissue engineering applications. J. Tissue Eng. Regen. Med. 5, 421–436.

Perchyonok, V.T., Zhang, S.M., Oberholzer, T., 2012a. Novel melatonin-chitosan hydrogels as suitable oral bio-drug delivery systems to fight oral mucositis: synergy of antioxidants and bioactives in action. Curr. Org. Chem. 16, 2430–2436.

Perchyonok, V.T., Zhang, S.M., Oberholzer, T., 2012b. Alternative chitosan based drug delivery system to fight oral mucositis: synergy of conventional and bioactives towards the optimal solution. Curr. Nanosci. 8, 541–547.

Pirraco, R.P., Melo-Ferreira, B., Santos, T.C., et al., 2013. Adipose stem cell-derived osteoblasts sustain the functionality of endothelial progenitors from the mononuclear fraction of umbilical cord blood. Acta Biomater. 9, 5234–5242.

Popa, E.G., Reis, R.L., Gomes, M.E., 2012. Chondrogenic phenotype of different cells encapsulated in κ-carrageenan hydrogels for cartilage regeneration strategies. Biotechnol. Appl. Biochem. 59, 132–141.

Popa, E.G., Reis, R.L., Gomes, M.E., 2014. Seaweed polysaccharide-based hydrogels used for the regeneration of articular cartilage. Crit. Rev. Biotechnol. http://dx.doi.org/10.3109/07388551.2014.889079.

Rajam, M., Pulavendran, S., Rose, C., Mandal, A.B., 2011. Chitosan nanoparticles as a dual growth factor delivery system for tissue engineering applications. Int. J. Pharm. 410, 145–152.

Rossi, S., Marciello, M., Ferrari, F., et al., 2012. Development of sponge-like dressings for mucosal/transmucosal drug delivery into vaginal cavity. Pharm. Dev. Technol. 17, 219–226.

Santo, V.E., Gomes, M.E., Mano, J.F., Reis, R.L., 2012. Chitosan–chondroitin sulphate nanoparticles for controlled delivery of platelet lysates in bone regenerative medicine. J. Tissue Eng. Regen. Med. 6, s47–s59.

Sakiyama-Elbert, S.E., Hubbell, J.A., 2001. Functional biomaterials: design of novel biomaterials. Annu. Rev. Mater. Res. 31, 183–201.

Sashiwa, H., Aiba, S.I., 2004. Chemically modified chitin and chitosan as biomaterials. Prog. Polym. Sci. 29, 887–908.

Scholtz, J., Van der Colff, J., Steenekramp, J., et al., 2014. More good news about polymeric plant- and algae-derived biomaterials in drug delivery systems. Curr. Drug Targets 15, 486–501.

Shukla, S.K., Mishra, A.K., Arotiba, O.A., Mamba, B.B., 2013. Chitosan-based nanomaterials: a state-of-the-art review. Int. J. Biol. Macromol. 59, 46–58.

Silva, T.H., Duarte, A.R.C., Moreira-Silva, J., et al., 2012a. Biomaterials from marine-origin biopolymers. In: Mano, J.F. (Ed.), Biomimetic Approaches for Biomaterials Development. Wiley-VCH, Weinheim, pp. 3–24.

Silva, T.H., Alves, A., Ferreira, B.M., et al., 2012b. Materials of marine origin: a review on polymers and ceramics of biomedical interest. Int. Mater. Rev. 57, 276–306.

Silva, T.H., Moreira-Silva, J., Marques, A.L.P., et al., 2014. Marine origin collagens and its potential applications. Mar. Drugs 12, 5881–5901.

Szarpak, A., Pignot-Paintrand, I., Nicolas, C., et al., 2008. Multilayer assembly of hyaluronic acid/poly(allylamine): control of the buildup for the production of hollow capsules. Langmuir 24, 9767–9774.

Tabata, Y., Ikada, Y., 1998. Protein release from gelatin matrices. Adv. Drug Deliv. Rev. 31, 287–301.

Temenoff, J.S., Mikos, A.G., 2000. Review: tissue engineering for regeneration of articular cartilage. Biomaterials 21, 431–440.

Tonnesen, H.H., Karlsen, J., 2002. Alginate in drug delivery systems. Drug Dev. Ind. Pharm. 28, 621–630.

Vasi, A.M., Popa, M.I., Butnaru, M., et al., 2014. Chemical functionalization of hyaluronic acid for drug delivery applications. Mater. Sci. Eng. C Mater. Biol. Appl. 38, 177–185.

Vieira, A.P., Badshah, S., Airoldi, C., 2013. Ibuprofen-loaded chitosan and chemically modified chitosans – release features from tablet and film forms. Int. J. Biol. Mol. 52, 107–115.

Wang, Y.C., Ho, C.C., 2004. Micropatterning of proteins and mammalian cells on biomaterials. FASEB J. 18, 525–527.

Wang, D.A., Varghese, S., Sharma, B., et al., 2007. Multifunctional chondroitin sulphate for cartilage tissue-biomaterial integration. Nat. Mater. 6, 385–392.

Woitiski, C.B., Veiga, F., Ribeiro, A., Neufeld, R., 2009. Design for optimization of nanoparticles integrating biomaterials for orally dosed insulin. Eur. J. Pharm. Biopharm. 73, 25–33.

Zang, M.Q., Zhang, Q.X., Davis, G., et al., 2011. Perichondrium directed cartilage formation in silk fibroin and chitosan blend scaffolds for tracheal transplantation. Acta Biomater. 7, 3422–3431.

Zuo, P.P., Feng, H.F., Xu, Z.Z., et al., 2013. Fabrication of biocompatible and mechanically reinforced graphene oxide-chitosan nanocomposite films. Chem. Centr. J. 7, 39.

Anti-inflammatory drug delivery systems using marine products

9

S.P. Victor, C.P. Sharma

9.1 Introduction

The oceans are home to around 90% of the world's living biomass, and the marine environment is an extraordinary reservoir of bioactive natural products. These compounds exhibit multiple pharmacological properties primarily due to their unique chemical and structural features, not found in terrestrial natural products (Mayer and Gustafson, 2003). The fact remains that these marine organisms have evolved both physiologically and biochemically to help them survive their hostile home environment. The secondary metabolites of these marine organisms, which are composed of small molecules, exhibit significant pharmacological effects like immunomodulation, anti-inflammatory, antibacterial and antiviral activities. Relatively low octanol–water partition coefficient, rotatable bonds and stereogenic centers are some of the chemical properties that these molecules possess, which make them favorable as structures for drug discovery (Mayer and Gustafson, 2004). Marine organisms are therefore considered as treasures that remain a relatively unexplored source for novel bioactive compounds that could eventually be developed into therapeutics. Marine products and chemicals have been reported to possess anthelmintic, antibacterial, anticoagulant, antifungal, antimalarial, antiprotozoal, antituberculosis, and antiviral properties. There are numerous publications describing marine compounds affecting the cardiovascular, immune and nervous systems, as well as those with anti-inflammatory effects and activity toward cellular and molecular targets (Mayer and Gustafson, 2006). This chapter discusses the anti-inflammatory properties of reported marine organisms, their chemistry, mechanisms of action and pharmacological activity.

9.2 Marine compounds with anti-inflammatory activity

Marine compounds obtained from different marine sources including sponges, molluscs, bryozoans, sea combs, algae, echinoderms, ascidians and soft corals are believed to be the source of potential candidate compounds for treating inflammatory diseases. Among the prominent marine compounds exerting exceptional anti-inflammatory activities are manoalide, γ-hydroxybutenolide isolated from the marine sponge *Luffariella variabilis*, pseudopterosin E, a tricyclic diterpene glycoside isolated from the Caribbean gorgonian *Pseudopterogorgia elisabethae* and cacospongionolide E (Wen et al., 2010), with a pyranofuranone moiety analogous to that of manoalide, isolated from the sponge *Fasciospongia cavernosa* (Lin et al., 2011). Astaxanthin,

Functional Marine Biomaterials. http://dx.doi.org/10.1016/B978-1-78242-086-6.00009-1
Copyright © 2015 Elsevier Ltd. All rights reserved.

the main carotenoid pigment present in the marine world of algae, has exhibited anti-inflammatory activities by suppressing the activation of NF-κB. It inhibits the production of pro-inflammatory mediators such as tumor necrosis factor (TNF)-α and IL-1β (Hussein et al., 2006). Hymenialdisine alkaloids isolated from marine sponges, such as *Acanthella aurantiaca* and *Stylissa massa*, interfere with NF-κB and decrease the production of IL-8 and IL-1β, thereby inducing potential anti-inflammatory properties (Jean et al., 2009). The marine compound sinularin from the soft coral *Sinularia flexibilis* has been reported to have anticancer activity against the human epidermoid carcinoma cell line (KB) and the murine P388 lymphocytic leukemia cell line (PS) (Weinheimer et al., 1977). The anti-inflammatory activity of sinularin was observed when its oral administration reduced carrageenan-induced paw edema 3h after carrageenan injection and after adjuvant administration inhibited adjuvant-induced paw swelling (periarthritis model) over a 21-day period (Buckle et al., 1980). Similarly, it has been reported that sinularin, isolated from the same genus but from a different species of soft coral, *Sinularia querciformis* significantly inhibited upregulation of proinflammatory proteins, particularly inducible nitric oxide synthase (iNOS) and cyclooxygenase-2 (COX-2), in lipopolysaccharide (LPS)-stimulated murine macrophage cells (Salvemini et al., 1996). New anti-inflammatory cembranolides have recently been identified from the soft coral *Lobophytum*. These cembranolides inhibited the expression of iNOS and COX octocorals (*Coelenterata*), have been well recognized as a rich source of hydroxy sterols and their polyoxygenated analogs (Cheng et al., 2009). Some of these compounds have also been shown to exhibit *in vitro* cytotoxic activities toward several cancer cell lines.

A study on *Sinularia gibberosa* reported in literature led to the isolation of new polyoxygenated sterols: gibberoketosterols along with the known steroids methylenecholest-5-*en*-3-ol. These cholestane-type sterols possessing a 22, 24(28)-conjugated diene in the side chain or a 5R, 6R-epoxide in the B-ring were isolated for the first time from marine sources. The newly isolated compound showed significant inhibition against the up-regulation of the pro-inflammatory iNOS and COX-2 proteins of LPS-stimulated RAW264.7 macrophage cells. The symbiotic marine dinoflagellate *Symbiodinium*, a type of zooxanthellae, is found in a wide range of marine invertebrates and produces several super-carbon-chain compounds, such as zooxanthellatoxins and zooxanthellamides. Two amphoteric iminium metabolites, symbioimine and neosymbioimine, were isolated from a cultivated symbiotic marine dinoflagellate *Symbiodinium*. They have a characteristic 6,6,6-tricyclic iminium ring structure and an aryl sulfate moiety (D'Orazio et al., 2012). The compound symbioimine inhibited the differentiation of RAW264 cells into osteoclasts and significantly inhibited COX-2 activity. Further, the Red alga *Gracilaria verrucosa* has been shown to inhibit LPS-induced nitric oxide (NO) production, TNF-α and IL-6 (Dang et al., 2008). The lipid extract of blue green algae *Nostoc* suppressed the expression of pro-inflammatory mediators, such as TNF-α, COX-2, IL-6, and IL-1β, by inhibiting the activation of NF-κB pathway linked to inflammation (Park et al., 2008).

Inflammation is a defensive response of vascular tissues to harmful stimuli such as injury, pathogens, or irritants. These responses are essential for humans as a defense mechanism in combating infections and promoting restoration and healing. However, deregulated inflammation results in a large number of diseases like autoimmune diseases, allergies, rheumatoid arthritis, atherosclerosis, Alzheimer's, asthma, multiple sclerosis, and inflammatory bowel diseases. The three major groups of drugs used in the treatment of inflammatory diseases are corticosteroids, nonsteroidal anti-inflammatory drugs (NSAIDs), and disease-modifying antirheumatic drugs, which include biological agents.

Inflammation involves a multifactorial network of chemical signals to mediate their action and is controlled by a group of substances called chemical mediators. The endogenous chemical mediators mainly consist of vasoactive amines, cytokines, complement components, eicosanoids, platelet-activating factor (PAF), and neuropeptides among which eicosanoids play a critical role in every step of inflammation. Eicosanoids, which mainly consist of prostaglandins, prostacyclins, thromboxanes, and leukotrienes, are a family of oxygenated fatty acid metabolized by cyclooxygenases (COXs) and lipoxygenases (LOXs) from arachidonic acid (Teixeira et al., 2003). The primary anti-inflammatory targets include COX-1 and -2 enzymes, cytokines such as TNF-α and interleukins (IL-1β, IL-6), transcription factors and several more. TNF-α and ILs are intercellular signal proteins that are released by immune cells and have numerous functions in the promotion and resolution of inflammation. They have been identified to play a central role in the pathogenesis of many inflammation diseases like asthma and rheumatoid arthritis. The transcription factor NF-κB regulates the expression of several genes involved in the activation of inflammation and plays a major role in the pathogenesis of inflammatory bowel and rheumatic diseases, respectively. It has been well established that the excessive production of proinflammatory mediators results in several inflammatory diseases. So, inhibition of the production of these mediators is an exciting option for treatment. Despite the extensive efforts invested in developing drugs that suppress the conversion of arachidonic acid into proinflammatory eicosanoids, the latter approach has been unsuccessful. Undesired side effects resulting from the lack of specificity of COX and LOX are responsible for the failure of the concept (Yedgar et al., 2000). As an alternative, the quest for inhibitors of phospholipases A2 (PLA2s), the enzymes that catalyze the hydrolysis of membrane phospholipids into arachidonic acid, has opened up a new research avenue in anti-inflammatory drug discovery (Burke and Dennis, 2009). Marine organisms like corals, snails, and sponges have been shown to inhibit the enzymatic activity of PLA2 and have active compounds that have exhibited anti-inflammatory activity through various molecular mechanisms.

9.3 Chemistry and pharmacological activity

Bolinaquinone and Petrosaspongiolide are two marine terpenes isolated from the sponges *Dysidea* and *Petrosaspongia nigra* (Busserolles et al., 2005). The hypothesis that its oral administration could inhibit inflammation and oxidative stress was investigated in an *in vivo* murine model of inflammatory bowel disease in humans. It was observed that both the compounds merosesquiterpene and sesterterpene inhibited neutrophilic infiltration, interleukin-1β, prostaglandin E2 levels, and COX-2 protein expression *in vivo*. The pharmacological activity of these compounds resulted in the inhibition of colonic inflammation *in vivo* suggesting that these compounds could be utilized for "protective strategies" against intestinal inflammatory diseases. Novel isozonarone and isozonarol derivatives (McNamara et al., 2005) obtained from the New Zealand sponge *Dysidea cristagalli* were found to inhibit free radical release of human neutrophil. *In vitro* studies with human neutrophils demonstrated a reduction of superoxide anion release that was found to be concentration dependent. It was explained on a mechanism hypothesized to involve the accumulation of the lipophilic sesquiterpene moiety in cell membranes, where it could interfere with superoxide anion production. Further, it was reported (Miyaoka et al., 2006) that if a compound possessing a γ-hydroxybutenolide moiety exhibits inhibitory activity against PLA2, it must have an appropriately sized hydrophobic substructure. They reported the pharmacology of PLA2 inhibitors by investigating two sesterterpenoids, cladocorans A and B, from the coral *Cladocora caespitosa*, which possess a γ-hydroxybutenolide moiety. Cladocorans A and B and their diastereomers were capable of inhibiting the activity of sPLA2 to an extent that almost equaled that of manoalide and were observed to potently inhibit secretory PLA2 ($IC50=0.8–1.9\,\mu M$). Another study (da Matta et al., 2011) went on to evaluate the anti-inflammatory activity of the methanolic, acetate, hexanic, and chloroform extracts obtained from marine algae *Caulerpa mexicana* and *Caulerpa sertularioides*, respectively. The effects of these extracts on peritoneal inflammation induced by carrageenan were studied on mice treated with test samples. The carrageenan-induced peritonitis test made it possible to produce an acute inflammatory response after 4 h in the peritoneal cavity of mice, with a large number of leukocytes in the exudates. All the extracts evaluated were able to significantly inhibit leukocyte migration into the peritoneal cavity in comparison with carrageenan. The results demonstrate that extracts from *Caulerpa* species elicit pronounced antinociceptive and anti-inflammatory activity against several nociceptive models.

Similarly, it has been observed that thromboxane B2 (TXB2) and superoxide anion are neuroinflammatory mediators that are involved in the pathogenesis of several neuroinflammatory disorders. The first experimental study that demonstrated the anti-neuroinflammatory properties of the indole-derived alkaloids manzamines isolated from marine sponges *Haliclona*, *Amphimedon,* and *Xestospongia* was carried out in 2005 (Mayer et al., 2005). They conducted a structure–activity relationship study to investigate manzamine's potent inhibition of superoxide anion ($IC50=0.1\,\mu M$) and TXB2 ($IC50=0.016\,\mu M$) release, respectively, to activated rat neonatal microglia. It was suggested that the activity of the compound was related to its solubility, ionic forms and changes such as saturation, oxidation of the β carboline and 8-membered amine ring, which played a critical role in the observed structure–activity relationship. The results demonstrate that manzamines, at *in vitro* concentrations that are nontoxic to *E. coli* LPS, activated rat neonatal microglia and potently modulated TXB2 and superoxide generation. Sarcophine, a cembranoid diterpene isolated from the Red Sea soft coral *Sarcophyton glaucum* is known to possess anti-inflammatory potential and inhibit the process of tumorigenesis. A study investigated the effects that sarcophine and its sulfur-containing derivatives obtained by targeting its epoxide ring had on the release of inflammatory mediators like TXB2 and superoxide anion. The effects of various concentrations of sarcophine and its semisynthetic derivatives were studied on highly malignant SA mammary epithelial cell proliferation in an activated rat neonatal microglia study. Sarcophine resulted in only minor reduction in SA mammary epithelial cell proliferation, whereas its derivatives were found to induce a significant reduction. Sarcophine and its derivatives were found to be cytostatic, but not cytotoxic, to neoplastic mammary epithelial cells grown in culture (Sawant et al., 2006). Further, the effects of three concentrations of sarcophine and its sulfur-containing analogs on anti-inflammatory activity were evaluated. The release of superoxide anion and TXB2 was monitored from LPS-activated rat neonatal brain macrophages. It was demonstrated that "targeting the epoxide ring of sarcophine" enhanced sarcophine's anti-inflammatory activity (Mayer et al., 1997).

Another active triterpene oligoglycosides, cucumarioside obtained from the Far-Eastern edible sea cucumber *Cucumaria japonica* is a source of 13 biologically active compounds (Aminin et al., 2006). All these compounds primarily consist of a carbohydrate chain with five monosaccharide units linked to C-3 of aglycons, which are represented by the lanostane 18(20)-lactones. These glycosides further contain one, two, or three sulfate groups bonded with sugars. Cucumariosides have been shown to significantly increase animal resistance against bacterial infections elicited by various pathogens. Moreover, it has also been found to stimulate phagocytosis, IL-6, and TNF-α production in lymphocytes, ROS formation, increase the number of antibody-producing cells and amplify the expression of several cell-surface molecules that were preliminary cultured with hydrocortisone. An increase in macrophage phagocytic activity was observed, which was found to double within 2 weeks with an ip injection of a mixture of cucumariosides at a minimal dose of 0.03 μg/mouse. This influence of Cumaside on mouse macrophages in low doses was accompanied by a two-fold stimulation of lysosomal activity. It has further been demonstrated that Cumaside causes a dose-dependent stimulation of humoral immune response in *in vivo* experiments and increased the number of antibody plaque-forming cells. Similarly, Cumaside causes a significant dose-dependent increase of TNF-α production in blood mononuclear cells of healthy volunteers and also shows the tendency to stimulate IL-6 release. At the same time, it has been observed that it did not affect the delayed-type hypersensitivity, proliferative activity of lymphocytes, the release of cytokine IFNγ and IL12p70 and did not affect cytotoxic activity of human NK cells. The ability of Cumaside to restore cell-surface molecules on lymphocytes with an inhibited metabolism could be beneficial in correction of depressed cell immunity. It also elicits stimulatory action on absorption of *S. aureus* both by monocytes and neutrophils of human peripheral blood. This work went on to reveal that Cumaside expressed evident immunostimulatory properties at low doses. This property was partly attributed to the action of TNF whose production was induced by Cumaside, and also to IL-6 cytokine induction which are believed to be the key processes in the molecular mechanism of Cumaside action.

It was also demonstrated (Mandeau et al., 2005) that the ethanolic extract of the sponge *Euryspongia* contained petrosterol and was shown to display anti-inflammatory activity. This petrosterol on partitioning and further chromatography yielded compounds 3-hydroxy-24-norchol-5-*en*-23-oic acid and 3β-hydroxy-26-norcampest-5-*en*-25-oic acid. The sterols were evaluated for their action on the metabolism of arachidonic acid and eicosanoids in the human keratinocyte cell line HaCaT. The effects of the steroids on calcimycin (A23187)-triggered release of metabolites of the COX pathway were evaluated. These compounds were further monitored for their anti-inflammatory activity against 6-keto-prostaglandin F1 release in a human keratinocyte cell line HaCaT. The steroid reduced 6KPGF1α production by human keratinocytes by 41% at a concentration of 10 μg/mL. Three new cholestane-type sterols were isolated for the first time from the Formosan soft coral *S. gibberosa* (Ahmed et al., 2006). These cholestane-type sterols possessing a 22,24(28)-conjugated diene in the side chain or a 5R,6R-epoxide in the B-ring showed significant inhibition against the up-regulation of the pro-inflammatory iNOS and COX-2 proteins of LPS-stimulated RAW264.7 macrophage cells. Among the isolated sterols, gibberoepoxysterol reduced the levels of the iNOS and COX-2 proteins at a concentration of 10 μM to 44.5% and 68.3%, respectively, relative to the control cells stimulated with LPS. However, the same concentration of the sterol gibberoketosterol B did not produce any inhibition of LPS-induced iNOS and COX-2 expression. The toxicity of these sterols to RAW264.7 cells was also assessed by trypan blue staining. The cytotoxicity of these sterols toward a limited panel of cancer cell lines has also been evaluated. Cell survival was not changed by the presence of the sterols at the concentration of 10 μM. Both cell loss and observed actin decrease occurred at higher concentration of the sterols. A series of polyprenylated hydroquinones, quinones, and chromenols were isolated from the extracts of the marine sponge *Ircinia spinosula* and the brown alga *Taonia atomaria*. Their anti-inflammatory activities in terms of leukotriene formation were evaluated in an *in vitro* assay with pork leukocytes. The new hydroxylated compounds, 2[28-hydroxy] heptaprenyl-1,4-hydroquinone and (2-[(2E,6E,10E,14E,18Z,22E)-19-(hydroxymethyl)-3,7,11,15,23,27-hexamethyloctacosa-2,6,10,14,18,22,26-heptaen-1-yl]benzene-1,4-diol), the known tetraprenyl benzoquinone sargaquinone, and the known polyprenyl chromenols exhibited the highest anti-inflammatory activities, with IC50 values of 1.9. Compounds structurally resembling the highly active tocopherol antioxidants would be expected to show greater anti-inflammatory activities than the respective hydroquinones. Nevertheless, the results obtained corroborate that the activity of hydroquinone is higher than that of the respective chromenol, suggesting that scavenging capacity alone is not the determining factor for the inhibition of 5-LOX. The authors (Tziveleka et al., 2005) presented that chromenol's potent inhibition of leukotriene B4 generation by stimulated porcine leukocytes (IC50 = 1.9 μM) was related to the "absence of a side chain OH group as well as the reduced number of prenyl moieties" on the sponge metabolite.

Three sesquiterpenoids with unprecedented skeletons, paralemnanone, isoparalemnanone, and paralemnanol, were isolated (Huang et al., 2006) from the Formosan soft coral *Paralemnalia thyrsoides* that significantly inhibited inflammatory iNOS protein expression (70% at 10 μM) in activated RAW264.7 cells. The inhibition of these metabolites toward the LPS-induced proinflammatory proteins (iNOS and COX-2) expression was also evaluated. In this assay, stimulation of the RAW264.7 cells with LPS resulted in accumulation of the proinflammatory iNOS and COX-2 proteins by immunoblot analysis. Both paralemnanone and isoparalemnanon at a concentration of 10 μM could reduce the levels of the iNOS to 48.7 ± 11.2% and 70.6 ± 3.8%, respectively, and COX-2 to 73 ± 3.1% and 68.5 ± 10.1%, respectively, in comparison with those of the control cells stimulation with LPS (100% for both iNOS and COX-2). Metabolite paralemnanol did not inhibit the COX-2 expression (99.7 ± 10.4%), but could reduce iNOS expression (66 ± 4.6%) by LPS treatment. It has also been reported (Sugiura et al., 2006) that a phlorofucofuroeckol-B from an edible Japanese marine brown alga, *Eisenia arborea*, inhibited histamine release (IC50 = 7.8 μM) from a rat basophilic leukemia in a concentration-dependent manner, an observation which compared favorably with a clinically used antihistamine Tranilast (IC50 = 46.6 μM). Two amphoteric iminium metabolites, symbioimine and neosymbioimine, were isolated from a cultivated symbiotic marine dinoflagellate *Symbiodinium*. The compounds had a characteristic 6,6,6-tricyclic iminium ring structure and an aryl sulfate moiety. The plausible biogenetic pathway of 1 and 2 could be explained by an intramolecular Diels–Alder reaction followed by imine cyclization. Symbioimine inhibited the differentiation of RAW264 cells into osteoclasts (EC50 = 44.1 M) and significantly inhibited COX-2 activity at 10 μM. Symbioimine inhibited osteoclastogenesis in the murine monocytic cell line RAW264, which can differentiate into osteoclasts following treatment with receptor activator of nuclear factor-kB ligand (RANKL) (EC50 = 44 μg/mL). RANKL induces the formation of osteoclast-like multinucleated cells in cultures of bone marrow cells. Symbioimine inhibited an increase in RANKL-induced tartrate-resistant acid phosphatase (TRAP) activity in preosteoclast cells. However, it did not affect cell viability even at 100 μg/mL. The authors (Kita et al., 2005) further propose that symbioimine might become a useful lead to develop new NSAIDs. It could also serve as a potential antiresorptive drug for the prevention and treatment of osteoporosis in postmenopausal women.

Marine PLA2 inhibitors have also been isolated and they are primarily terpenoids isolated from sponges, nudibranchs, and algae. Among all the reported terpenoids, sesterterpenes have an outstanding potential as an anti-inflammatory compound. The sesterterpene manoalide isolated for the first time from the sponge *L. variabilis* (De Silva and Scheuer, 1980) was the first marine natural product reported as a PLA2 inhibitor. Manoalide has been shown to inhibit PLA2 (IC50 = 10 M) and phospholipase C (Matsumaga and D'Auria, 2000). It is one of the most widely investigated marine PLA2 antagonists. Mechanistic studies have revealed that its inhibitory activity results from the irreversible binding of two of the compound's masked aldehyde groups (the hydroxydihydropyran ring and the hydroxybutenolide ring) to lysine residues at the active site of PLA2. Manoalide had been licensed to Allergan Pharmaceuticals and had reached Phase II clinical trials for use as a topical antipsoriatic. However, its development was discontinued due to problems associated with formulation. Several analogs of the molecule manoalide have been isolated from sponges belonging to the genus *Luffariella*, as well as from other sponges. Manoalide analogs isolated from nudibranchs of the *Chromodoris* genus, which include luffariellins, deoxymanoalide, secomanoalide, luffariellolide, and deoxysecomanoalide, are all reduced (deoxy) counterparts of spongean manoalide analogs, and their PLA2 inhibitory activity is 10-fold weaker than the ones observed in the sponges.

The diterpenes gracilin A, aplyroseol, and 12-acetoxytetrahydroaplysulphurin isolated from *Aplysilla* sp. sponges (Potts et al., 1992), and dendrillolide A and norrisolide isolated from the sponge *Dendrilla* sp. inhibit bee venom sPLA2 with IC50 values around 5 M. They all contain a masked 1,4-dialdehyde function, which has been suggested to play a key role in their bioactivity (Nirmal et al., 2008). The meroditerpne epitaondiol isolated from the brown alga *Stypopodium flabelliforme* inhibits TXB2 production by potently inhibiting human sPLA2 (IC50 = 3.8 M) (König and Wright, 1996). The tetra- and bicyclic diterpenes phomactins A–C isolated from the marine fungus *Phoma* sp. are potent PAF antagonists. While the precise mode of action remains poorly understood, it is likely that these three compounds may act as PAF antagonists by inhibiting PLA2s or by activating (Lp)PLA2 (Sugano et al., 1991). The arabidose-containing diterpene fuscoside B isolated from the gorgonian *Eunicea fusca* has not been reported as a PLA2 inhibitor, but it has been shown to inhibit the conversion of arachidonic acid to LTB4 by inhibiting 5-LOX (IC50 = 18 M) (Dunlap et al., 2007).

Sesquiterpenoids quinine and ilimaquinone isolated from the sponge *Hippiospongia metachromia* has also been shown to inhibit PLA2. Lactone cavernolide isolated from the sponge *F. cavernosa* inhibits sPLA2 activation (IC50 = 8.8 M), as well as iNOS and COX-2 gene expressions (Mayer et al., 1993). The anti-psoriasis sesquiterpene hydroquinone avarol and the sesquiterpene quinones avarone and dysidine isolated from the sponge *Dysidea avara* inhibit PLA2 activity and PGE2 release in keratinocytes and in monocytes (IC50 = 2 M). Furthermore, avarol has also been shown to reduce eicosanoid release and ROS generation in stimulated leukocytes. Dysidiotronic acid isolated from *Dysidea* also inhibits sPLA2 (IC50 = 2.6 M) (Nakao and Fusetani, 2007).

Other PLA2-inhibiting sesterterpenes isolated from various marine sponges include cacospongiolide B (IC50 = 300 nM against human and bee venom sPLA2), cyclolinteinone (IC50 = 25 M against bee venom sPLA2), petrosaspongiolide M (IC50 = 1.6 M against human sPLA2; 0.6 M against bee venom PLA2), scalaradial (IC50 = 1.6 nM against bee sPLA2 and cPLA2), aplyolide (IC50 = 10.5 M against human sPLA2), and palauolide (IC50 = 0.8 g/mL against bee venom sPLA2) (Ebada et al., 2010). Molecular modeling studies on petrosaspongiolide M reveal that PLA2 inhibition occurs by a noncovalent recognition between petrosaspongiolide M and the enzyme, followed by a nucleophilic attack. The pyridinium alkaloids spongidines A–D isolated from the sponge *Spongia* sp. inhibit human sPLA2 (IC50 = 10) due to the presence of hydroxybutenolide or masked 1,4-dialdehyde group (Potts et al., 1992). One of the most recently discovered marine PLA2 inhibitors shown to inhibit bacterial PLA2 (IC50 = 2 g/mL) is the methoxylated fatty acid 7-methoxy-9-methylhexadeca-4,8-dienoic acid obtained from the brown alga *Ishige okamurae* (Cho et al., 2008). Recently (Lins et al., 2012), the proinflammatory activity of SVHV, an alginate isolated from *Sargassum vulgare,* was investigated using models of paw edema, mast cells degranulation and neutrophil migration *in vivo*. SVHV induced a dose-dependent paw edema, with a peak at 2 h, associated with an increased myeloperoxidase activity and production of TNF-α and IL-1β. Pharmacological modulators, remarkably dexamethasone and indomethacin, inhibited the edema. SVHV (1.0 mg) also led to a significant induction of neutrophil migration in the peritoneal cavity of rats. This neutrophil migration was significantly reduced by peritoneal resident macrophages depletion, but was not affected by the depletion of mast cells. The results indicate that SVHV has proinflammatory activity that depends on the activation of the macrophage-resident cells.

9.4 Conclusion and future scope

Several reviews covering both general and specific subject areas of marine pharmacology are available in literature. Given the critical role of inflammation in diseases, identifying and developing novel anti-inflammatory drug candidates is of great importance in drug discovery. Though tremendous efforts have been invested in developing drugs that suppress the conversion of arachidonic acid into proinflammatory eicosanoids, the approach has not been favorable. As an alternative approach, the quest for inhibitors of PLA2s has opened up new venues in anti-inflammatory drug discovery. PLA2 plays a critical role in inflammation and is considered as an interesting target for anti-inflammatory drugs. Although the quest for novel marine PLA2 inhibitors decreased during the last decade, the last few years have resulted in significant progress in the field. However, one of the major challenges faced by researchers is that they need to identify compounds that express high levels of specificity to the PLA2 that they are inhibiting. The development of a thorough understanding of the chemical and biological properties of the various types of PLAs, and their mechanism of action has to be recognized and understood. Among the marine natural products included in this chapter, bolinaquinone and manoalide have been shown to potently inhibit PLA2s, but in a nonspecific manner. Further, specific anti-inflammatory inhibitors are more likely to be bioactive at lower concentrations than nonspecific inhibitors, and they are less prone to induce undesired side effects. Additionally, significant progress can be made by a better understanding of the biochemistry and biology of PLA2s. We can therefore expect a high number of novel, highly promising PLA2 inhibitors to be developed from marine sources, which will play a critical role in paving the future for anti-inflammatory drug delivery using marine products.

References

Ahmed, A.F., Hsieh, Y.T., Wen, Z.H., Wu, Y.C., Sheu, J.H., 2006. Polyoxygenated sterols from the Formosan soft coral *Sinularia gibberosa*. J. Nat. Prod. 69, 1275–1279.

Aminin, D.L., Pinegin, B.V., Pichugina, L.V., Zaporozhets, T.S., Agafonova, I.G., Boguslavski, V.M., Silchenko, A.S., Avilov, S.A., Stonik, V.A., 2006. Immunomodulatory properties of Cumaside. Int. Immunopharmacol. 6, 1070–1082.

Buckle, P.J., Baldo, B.A., Taylor, K.M., 1980. The anti-inflammatory activity of marine natural products – 6-n-tridecylsalicylic acid, flexibilide and dendalone 3-hydroxybutyrate. Agents Actions 10, 361–367.

Burke, J.E., Dennis, E.A., 2009. Phospholipase A2 structure/function, mechanism, and signaling. J. Lipid Res. 50, S237–S242.

Busserolles, J., Paya, M., D'Auria, M.V., Gomez-Paloma, L., Alcaraz, M.J., 2005. Protection against 2,4,6-trinitrobenzenesulphonic acid-induced colonic inflammation in mice by the marine products bolinaquinone and petrosaspongiolide M. Biochem. Pharmacol. 69, 1433–1440.

Cheng, S.Y., Wen, Z.H., Wang, S.K., Chiou, S.F., Hsu, C.H., Dai, C.F., Duh, C.Y., 2009. Anti-inflammatory cembranolides from the soft coral *Lobophytum durum*. Bioorg. Med. Chem. 17, 3763–3769.

Cho, J.Y., Gyawali, Y.P., Ahn, S.H., Khan, N.A., Knong, I.S., Hong, Y.K., 2008. A methoxylated fatty acid isolated from the brown seaweed *Ishige okamurae* inhibits bacterial phospholipase A2. Phytother. Res. 22, 1070–1074.

da Matta, C.B., de Souza, E.T., de Queiroz, A.C., de Lira, D.P., de Araújo, M.V., Cavalcante-Silva, L.H., de Miranda, G.E., de Araújo-Júnior, J.X., Barbosa-Filho, J.M., de Oliveira Santos, B.V., Alexandre-Moreira, M.S., 2011. Antinociceptive and anti-inflammatory activity from algae of the genus *Caulerpa*. Mar. Drugs 9, 307–318.

Dang, H.T., Lee, H.J., Yoo, E.S., Shinde, P.B., Lee, Y.M., Hong, J., Kim, D.K., Jung, J.H., 2008. Anti-inflammatory constituents of the red alga *Gracilaria verrucosa* and their synthetic analogues. J. Nat. Prod. 71, 232–240.

De Silva, E.D., Scheuer, P.J., 1980. Manoalide, an antibiotic sesterterpenoid from the marine sponge *Luffariella variabilis*. Tetrahedron 21, 1611–1614.

D'Orazio, N., Gammone, M.A., Gemello, E., De Girolamo, M., Cusenza, S., Riccioni, G., 2012. Marine bioactives: pharmacological properties and potential applications against inflammatory diseases. Mar. Drugs 10, 812–833.

Dunlap, W.C., Battershill, C.N., Liptrot, C.H., Cobb, R.E., Bourne, D.G., Jaspars, M., Long, P.F., Newman, D.J., 2007. Biomedicinals from the phytosymbionts of marine invertebrates: a molecular approach. Methods 42, 358–376.

Ebada, S.S., Lin, W.H., Proksch, P., 2010. Bioactive sesterterpenes and triterpenes from marine sponges: occurrence and pharmacological significance. Mar. Drugs 8, 313–346.

Huang, H.C., Wen, Z.H., Chao, C.H., Ahmed, A.F., Chiang, M.Y., Kuo, Y.H., Hsu, C.H., Sheu, J.H., 2006. Novel sesquiterpenoids from the Formosan soft coral *Paralemnalia thyrsoides*. Tetrahedron Lett. 47, 8751–8755.

Hussein, G., Sankawa, U., Goto, H., Matsumoto, K., Watanabe, H., 2006. Astaxanthin, a carotenoid with potential in human health and nutrition. J. Nat. Prod. 69, 443–449.

Jean, Y.H., Chen, W.F., Sung, C.S., Duh, C.Y., Huang, S.Y., Lin, C.S., Tai, M.H., Tzeng, S.F., Wen, Z.H., 2009. Capnellene, a natural marine compound derived from soft coral, attenuates chronic constriction injury-induced neuropathic pain in rats. Br. J. Pharmacol. 158, 713–725.

Kita, M., Ohishi, N., Washida, K., Kondo, M., Koyama, T., Yamada, K., Uemura, D., 2005. Symbioimine and neosymbioimine, amphoteric iminium metabolites from the symbiotic marine dinoflagellate *Symbiodinium* sp. Bioorg. Med. Chem. 13, 5253–5258.

König, G.M., Wright, A.D., 1996. Marine natural products research: current directions and future potential. Planta Med. 62, 193–211.

Lin, Y.C., Huang, S.Y., Jean, Y.H., Chen, W.F., Sung, C.S., Kao, E.S., Wang, H.M., Chakraborty, C., Duh, C.Y., Wen, Z.H., 2011. Intrathecal lemnalol, a natural marine compound obtained from formosan soft coral, attenuates nociceptive responses and the activity of spinal glial cells in neuropathic rats. Behav. Pharmacol. 22, 739–750.

Lins, K.O.A.L., Vale, M.L., Ribeiro, R.A., Costa-Lotufo, L.V., 2012. Proinflammatory activity of an alginate isolated from *Sargassum vulgare*. Carbohydr. Polym. 92, 414–420.

Mandeau, A., Debitus, C., Aries, M.F., David, B., 2005. Isolation and absolute configuration of new bioactive marine steroids from Euryspongia n. sp. Steroids 70, 873–878.

Matsumaga, M.V., D'Auria, M.V., 2000. Dysidotronic acid, a new and selective human phospholipase A2 inhibitor from the sponge *Dysidea* sp. Tetrahedron Lett. 41, 3257–3260.

Mayer, A.M.S., Gustafson, K.R., 2003. Marine pharmacology in 2000: antitumor and cytotoxic compounds. Int. J. Cancer 105, 291–299.

Mayer, A.M.S., Gustafson, K.R., 2004. Marine pharmacology in 2001–2: antitumor and cytotoxic compounds. Eur. J. Cancer 40, 2676–2704.

Mayer, A.M.S., Gustafson, K.R., 2006. Marine pharmacology in 2003–2004: anti-tumour and cytotoxic compounds. Eur. J. Cancer 42, 2241–2270.

Mayer, A.M.S., Paul, V.J., Fenical, W., Norris, J.N., de Carvalho, M.S., Jacobs, R.S., 1993. Phospholipase A2 inhibitors from marine algae. Hydrobiologia 261, 521–529.

Mayer, A.M.S., Oh, S., Presto, E., Glaser, K.B., Jacobson, P.B., 1997. LPS-primed rat brain microglia: a convenient in vitro model to search for anti-inflammatory marine natural products. SHOCK 7, 49.

Mayer, A.M.S., Hall, M.L., Lynch, S.M., Gunasekera, S.P., Sennett, S.H., Pomponi, S.A., 2005. Differential modulation of microglia superoxide anion and thromboxane B2 generation by the marine manzamines. BMC Pharmacol. 5, 6.

McNamara, C.E., Larsen, L., Perry, N.B., Harper, J.L., Berridge, M.V., Chia, E.W., Kelly, M., Webb, V.L., 2005. Anti-inflammatory sesquiterpene-quinones from the New Zealand sponge *Dysidea* cf. *cristagalli*. J. Nat. Prod. 68, 1431–1433.

Miyaoka, H., Yamanishi, M., Mitome, H., 2006. PLA2 inhibitory activity of marine sesterterpenoids cladocorans, their diastereomers and analogues. Chem. Pharm. Bull. 54, 268–270.

Nakao, Y., Fusetani, N., 2007. Enzyme inhibitors from marine invertebrates. J. Nat. Prod. 70, 689–710.

Nirmal, N., Om Praha, G., Velmurugan, D., 2008. Modelling studies on phospholipase A2-inhibitor complexes. Indian J. Biochem. Biophys. 45, 256–262.

Park, Y.K., Rasmussen, H.E., Ehlers, S.J., Blobaum, K.R., Lu, F., Schlegal, V.L., Carr, T.P., Lee, J.Y., 2008. Repression of proinflammatory gene expression by lipid extract of Nostoc commune var sphaeroides Kutzing, a blue-green alga, via inhibition of nuclear factor-kappaB in RAW 264.7 macrophages. Nutr. Res. 28, 83–91.

Potts, B.C., Faulkner, D.J., Jacobs, R.S., 1992. Phospholipase A2 inhibitors from marine organisms. J. Nat. Prod. 55, 1701–1717.

Salvemini, D., Wang, Z.Q., Wyatt, P.S., Bourdon, D.M., Marino, M.H., Manning, P.T., Currie, M.G., 1996. Nitric oxide: a key mediator in the early and late phase of carrageenan-induced rat paw inflammation. Br. J. Pharmacol. 118, 829–838.

Sawant, S., Youssef, D., Mayer, A., Sylvester, P., Wali, V., Arant, M., El, S.K., 2006. Anticancer and anti-inflammatory sulfur-containing semisynthetic derivatives of sarcophine. Chem. Pharm. Bull. 54, 1119–1123.

Sugano, M., Sato, A., Iijima, Y., Oshima, T., Furuya, K., Kuwano, H., Hata, T., Hanzawa, H., 1991. Phomactin A: a novel PAF antagonist from a marine fungus *Phoma* sp. J. Am. Chem. Soc. 113, 5463–5464.

Sugiura, Y., Matsuda, K., Yamada, Y., Nishikawa, M., Shioya, K., Katsuzaki, H., Imai, K., Amano, H., 2006. Isolation of a new anti-allergic phlorotannin, phlorofucofuroeckol-B, from an edible brown alga, *Eisenia arborea*. Biosci. Biotechnol. Biochem. 70, 2807–2811.

Teixeira, C.F.P., Landucci, E.C.T., Antunes, E., Chacur, M., Cury, Y., 2003. Inflammatory effects of snake venom myotoxic phospholipases A2. Toxicon 42, 947–962.

Tziveleka, L.A., Abatis, D., Paulus, K., Bauer, R., Vagias, C., Roussis, V., 2005. Marine polyprenylated hydroquinones, quinones, and chromenols with inhibitory effects on leukotriene formation. Chem. Biodivers. 2, 901–909.

Weinheimer, A.J., Matson, J.A., Hossain, M.B., van der Helm, D., 1977. Marine anticancer agents: sinularin and dihydrosinularin, new cembranolides from the soft coral, *Sinularia flexibilis*. Tetrahedron Lett. 18, 2923–2926.

Wen, Z.H., Chao, C.H., Wu, M.H., Sheu, J.H., 2010. A neuroprotective sulfone of marine origin and the in vivo anti-inflammatory activity of an analogue. Eur. J. Med. Chem. 45, 5998–6004.

Yedgar, S., Lichtenberg, D., Schnitzer, E., 2000. Inhibition of phospholipase A2 as a therapeutic target. Biochim. Biophys. Acta 1488, 182–187.

Index

Note: Page numbers followed by *f* indicate figures and *t* indicate tables.

Printed in the United States
By Bookmasters